Computational Methods and GIS Applications in Social Science – Lab Manual

This lab manual is a companion to the third edition of the textbook *Computational Methods and GIS Applications in Social Science*. It uses the open-source platform KNIME to illustrate a step-by-step implementation of each case study in the book. KNIME is a workflow-based platform supporting visual programming and multiple scripting language such as R, Python, and Java. The intuitive, structural workflow not only helps students better understand the methodology of each case study in the book, but also enables them to easily replicate, transplant and expand the workflow for further exploration with new data or models. This lab manual could also be used as a GIS automation reference for advanced users in spatial analysis.

FEATURES

- The first hands-on, open-source KNIME lab manual written in tutorial style and focused on GIS applications in social science
- Includes 22 case studies from the United States and China that parallel the methods developed in the textbook
- Provides clear step-by-step explanations on how to use the open-source platform KNIME to understand basic and advanced analytical methods through real-life case studies
- Enables readers to easily replicate and expand their work with new data and models
- A valuable guide for students and practitioners worldwide engaged in efforts to develop GIS automation in spatial analysis

This lab manual is intended for upper-level undergraduate and graduate students taking courses in quantitative geography, spatial analysis, GIS applications in socioeconomic studies, GIS applications in business, and location theory, as well as researchers in the similar fields of geography, city and regional planning, sociology, and public administration.

Computational Methods and GIS Applications in Social Science – Lab Manual

Lingbo Liu and Fahui Wang

CRC Press
Taylor & Francis Group
Boca Raton London New York

CRC Press is an imprint of the
Taylor & Francis Group, an **informa** business

Designed cover image: © Lingbo Liu

First edition published 2024
by CRC Press
2385 NW Executive Center Drive, Suite 320, Boca Raton FL 33431

and by CRC Press
4 Park Square, Milton Park, Abingdon, Oxon, OX14 4RN

CRC Press is an imprint of Taylor & Francis Group, LLC

© 2024 Lingbo Liu and Fahui Wang

ISBN: 978-1-032-30243-0 (pbk)
ISBN: 978-1-003-30435-7 (ebk)

DOI: 10.1201/9781003304357

Typeset in Times
by Apex CoVantage, LLC

Access the Support Material: https://www.routledge.com/9781032266817

Contents

List of Figures

List of Tables

Authors

Lingbo Liu is a postdoctoral fellow at the Center for Geographic Analysis, Harvard University, leading the development of Geospatial Analytics Extension for KNIME. He was a Lecturer at the Department of Urban Planning, School of Urban Design, Wuhan University, from 2005 to 2022, and earned a PhD in digital urban administration and planning at Wuhan University in 2018. His research uses multi-source data and quantitative models to capture the spatiotemporal features of urban systems and provides decision support for public policy, sustainable urban planning, and design.

Fahui Wang is Associate Dean of the Pinkie Gordon Lane Graduate School and Cyril and Tutta Vetter Alumni Professor in the Department of Geography and Anthropology, Louisiana State University. He earned a BS in geography at Peking University, China, and an MA in economics and a PhD in city and regional planning at the Ohio State University. His research has revolved around the broad theme of spatially integrated computational social sciences, public policy and planning in geographic information systems. He is among the top 1% most-cited researchers in geography in the world.

1 Getting Started with KNIME and Its Geospatial Analytics Extension

This chapter introduces KNIME and its application in a case study. Section 1.1 introduces the installation of KNIME Analytics Platform and its user interface. Section 1.2 illustrates the installation and functionality of Geographic Analytics Extension for KNIME, the KNIME Python integration and the KNIME Interactive R Statistics Integration. Section 1.3 uses the KNIME Analytics Platform to implement Case Study 1 on analyzing the urban population density pattern in Baton Rouge, Louisiana. Several popular GIS tasks are practiced, such as attribute query, spatial query, attribute join, spatial join, projection, geometric attribute calculation, buffer, spatial overlay, dissolve, intersect, etc.

1.1 KNIME ANALYTICS PLATFORM

This section covers how to download and install KNIME Analytics Platform (AP) and illustrates its user interface layouts (modern UI and classical UI), general contents and functions.

Step 1. Downloading KNIME Analytics Platform: Go to *www.knime.com/* and click Download . On the web page of Download KNIME Analytics Platform, input your information (e.g., sign up for the newsletter; other updates are recommended but optional), and click Download to open the webpage of download links (Figure 1.1).

Download the proper version of KNIME AP installation files according to your computer Operation System (OS). Extract the archive or run the installer and start the executable file (under Windows, ⚠ 'knime.exe'; under Linux 'knime' and under Mac, 'knime. app'). For detailed instructions, please refer to the file readme.txt in the archive.

For better performance, users can revise the maximum RAM usage limitation in the file knime.ini under the installation folder. Open the knime.ini, and revise "-Xmx1024m" to "-Xmx8g". It increases the RAM limitation for KNIME AP from 1024 Megabytes (Mb) to 8 Gigabytes (Gb).

Step 2. Starting KNIME AP and designating workspace: Start KNIME AP. During initiating, users need to use *B*rowse button to specify a Workspace for KNIME AP in the dialog of KNIME Analytics Platform Launcher (Figure 1.2). The KNIME workspace is a folder on your local computer to store your KNIME workflows, node settings, and data produced. The workflows and data stored in your workspace are subsequently available through the KNIME Explorer in the upper left corner of KNIME AP. Users can run several KNIME AP programs simultaneously with different workspaces. The workspace should be on a disk with enough free space.

DOI: 10.1201/9781003304357-1

Windows

KNIME Analytics Platform for Windows (installer) *The installer adds an icon to the desktop and suggests suitable memory settings*	**Download** (462 MB)
KNIME Analytics Platform for Windows (self-extracting archive) *The self-extracting archive only creates a folder holding the KNIME installation*	**Download** (463 MB)
KNIME Analytics Platform for Windows (zip archive)	**Download** (552 MB)

Linux

KNIME Analytics Platform for Linux	**Download** (563 MB)

Mac

KNIME Analytics Platform for macOS x86_64 (Intel)	**Download** (536 MB)
KNIME Analytics Platform for macOS arm64 (Apple silicon)	**Download** (529 MB)

FIGURE 1.1   Download KNIME Analytics Platform

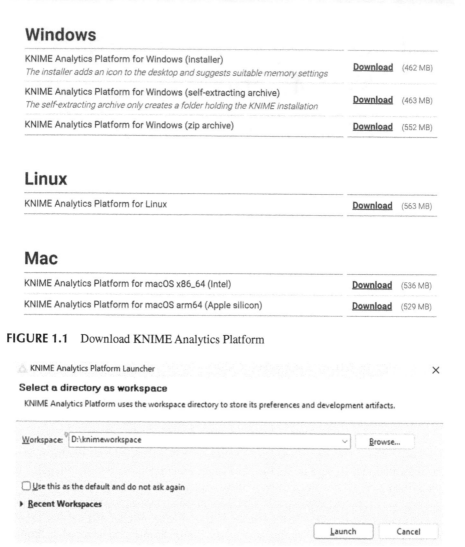

FIGURE 1.2   Dialog of KNIME Analytics Platform Launcher

Step 3. Optional: KNIME AP Modern UI interface: For KNIME AP 5.0 or a newer version, users can click Open KNIME Modern UI Preview. In the interface (Figure 1.3), click the black bar under Local space to enter the workspace (Local space). Either click Example Workflows to view the example workflows or click the button ✛ to create a new workflow.

For users who have registered in the KNIME community, click the button Sign in for My-KNIM-Hub to get the free cloud space provided by KNIME for storing and sharing workflows. It has public and private spaces for users to share and organize their workflows and components on the KNIME Hub. Workflows and components

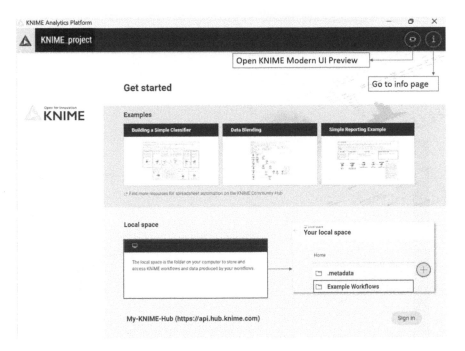

FIGURE 1.3 KNIME Modern UI Entry page

uploaded to public space are publicly available and therefore shared with the entire KNIME community. Workflows and data in the private space (maximum 1GB) are only visible to the account owner.

Figure 1.4 shows the interface of KNIME Modern UI with five main parts, such as ⊕ Node repository, ☰ Description, ⊕ Space explorer, Editor, and Preview. The nodes in the repository can be dragged to Editor, where users construct and edit the workflow for data analysis and visualization. After the node execution, users can click the node to view the result table in the Data Preview window. Description provides detailed information about the selected node, and Space explorer works as a file explorer to manage the workflows.

Step 4. KNIME AP classical user interface: Click the button ⟲ in the KNIME Modern user interface (UI) to switch back to the classic UI, which is adopted throughout all the case studies in this book.[1] After users become familiar with various nodes, switch back to the Modern UI. Alternatively, click the button ⓘ to open the information page and click Switch to alternate between the two UIs.

As shown in Figure 1.5, the classical interface mainly contains 10 components:

- Main Menu contains submenus of File, Edit, View, Node and Help. Under File, there are serval popular functions such as Import (Export) workflow, Switch workspace, Preference, Install KNIME extension, and Update KNIME. View helps us to control the interface layout.

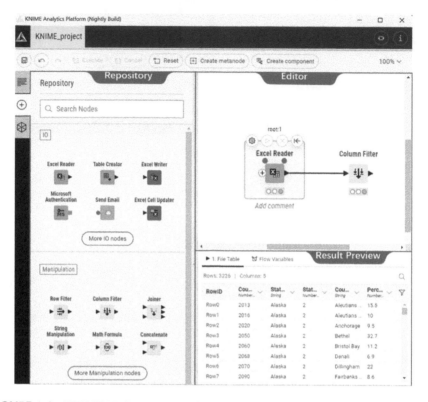

FIGURE 1.4   KNIME Modern User Interface

- KNIME Explorer provides an overview of the available workflows and workflow groups in the active KNIME workspaces, i.e., your local workspace as well as KNIME Servers.
- Workflow Coach lists node recommendations based on the workflows built by the wide community of KNIME users. It is inactive if you do not allow KNIME to collect your usage statistics.
- Node Repository includes all nodes available in core KNIME AP and in the extensions you have installed. The nodes are organized by categories, and you can also use the search box to find nodes.
- Workflow Editor is the canvas for editing the currently active workflow.
- Outline provides an overview of the currently active workflow.
- Console shows execution messages (e.g., what is going on under the hood), and Node Monitor shows the flow variable values or a preview of the output data of a selected node.
- Description is associated with the currently active workflow, or a node selected in the node repository or workflow editor. For a workflow, the first part is a general description, followed by tags and links to other resources related to the workflow. For a node, the first part is a general description, followed by the available setting options, and finally a list of input and output ports.

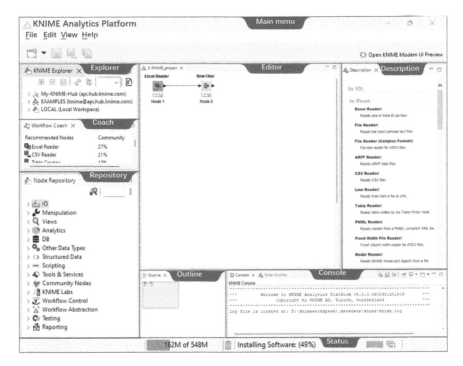

FIGURE 1.5 Classic interface of KNIME Analytics Platform

- KNIME Hub Search shows the search results in the KNIME Hub, and it is juxtaposed with Description tab.
- Status shows the status of RAM usage[2] or process of installing software.

1.2 KNIME EXTENSION FOR SPATIAL ANALYSIS

This section first covers how to install Geospatial Analytics Extension for KNIME. In advanced case studies, users need to access KNIME Python Integration and KNIME Interactive R Statistics Integration. One may skip steps 6–8 and come back for reference when needed.

The KNIME Python Integration contains nodes that allow the execution of Python 3 scripts in KNIME. It brings significant performance improvements over the legacy KNIME Python Integration and enables working with larger-than-memory data. This extension provides a Python installation with a small set of selected Python packages. Alternatively, the nodes can use a local Python installation.

The KNIME Interactive R Statistics Integration allows users to write and execute R scripts by interacting with an external R installation. This integration comprises a handful of nodes for users to write R scripts and execute them in KNIME AP. Besides providing the ability to read data from R into KNIME AP, these nodes allow users to manipulate data and create views using R, and apply models trained in R to the data.

Step 1. Installing KNIME Extensions: Start KNIME AP, go to File and click Install KNIME extensions . . . to bring up the dialog of Install (Figure 1.6). Under the box of Available software, (1) expand KNIME & Extensions, check KNIME Python Integration and KNIME Interactive R Statistics Integration, (2) expand KNIME Community Extension-Other, check Geospatial Analytics Extension for KNIME, and (3) click Next at the bottom. Check the information of Install Details and click Next. Review the information of Review License and accept the terms of the license agreements. Click Finish to install the extension. The installation status is shown at the bottom of KNIME AP (Figure 1.5).

After installation, restart KNIME AP to load the new extensions and their nodes.

Step 2. Configuring R environment: To use nodes in KNIME Interactive R Statistics Integration, one needs to download and install R. Go to the official website (*https://cran.r-project.org/*) to download R, e.g., the link for R 4.2.2 of Window version is *https://cran.r-project.org/bin/windows/base/old/4.2.2/R-4.2.2-win.exe*. In Windows OS, it is strongly suggested not to install R in the C drive, which may cause the default library unwritable.[3]

Then in KNIME AP, go to File > Preference > KNIME > R, use Browse to specify the installation path to R, and click Apply and Close (Figure 1.7).

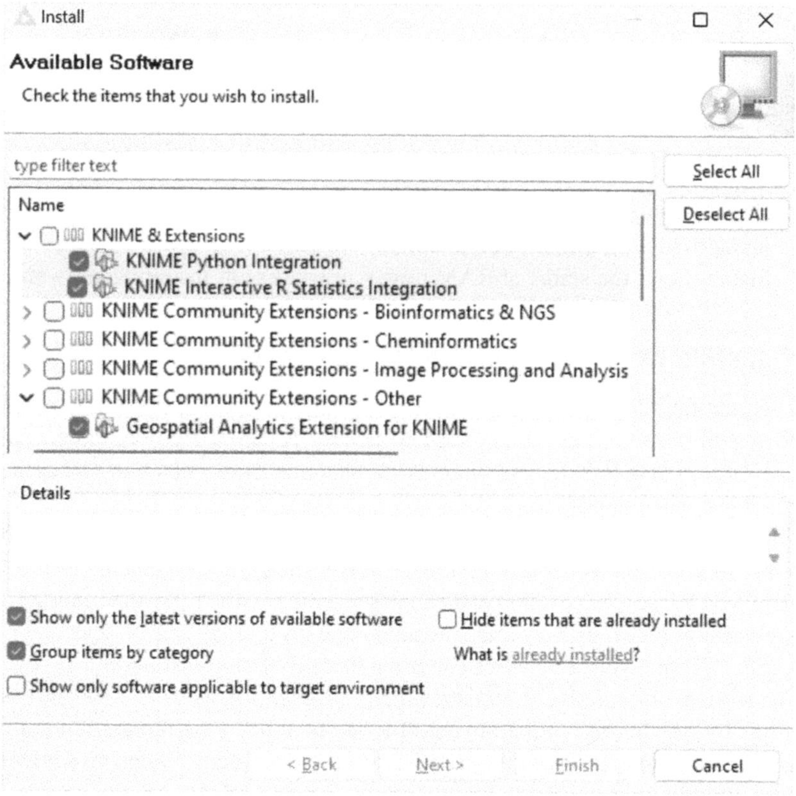

FIGURE 1.6   Interface of Available Software for installing KNIME extensions

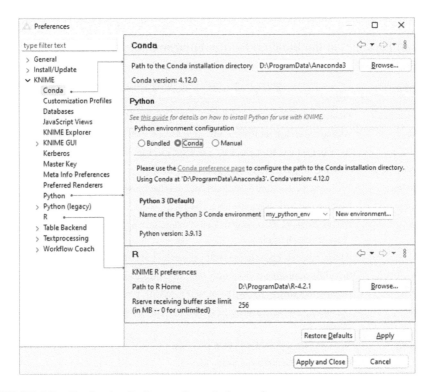

FIGURE 1.7 Configuring Preference for scripting environments

Rserve is an R package that allows other applications to talk to R using TCP/IP or local sockets sending binary requests. It requires different installation steps and different packages depending on the operating system. For Windows OS, from within R, run the command: `install.packages ('Rserve')`. For MacOS and Linux Ubuntu, run R from Terminal and type the command:

```
install.packages('Rserve',"http://rforge.net/",type="source")
```

For more information, visit *https://docs.knime.com/2019-12/r_installation_guide/index.html*.

Step 3. Optional: Understanding the bundled Python environment: The bundled Python environment lets us start Python Script node right away. This convenience allows us to use the nodes without installing, configuring or even knowing environments. Table 1.1 lists the packages included in the bundled Python environment.[4]

Step 4. Configuring Python environment: While the KNIME Python Integration comes with a bundled Python environment, users still need to install packages for specific geospatial functions (e.g., Anaconda) not included in the bundled environment.

Go to the official site (*www.anaconda.com*), download the suitable version according to the OS system, and create a new environment, e.g., named `my_python_env`.

TABLE 1.1
Built-in Packages in the Bundled Python Environment

Package Name	Function
beautifulsoup4	Parsing and navigating HTML and XML documents, often used for web scraping tasks.
cloudpickle	Pickling Python objects so that they can be saved and loaded across a network.
ipython	An interactive shell for Python, providing features such as tab completion and history.
jedi	Autocompletion and static analysis of Python code, often used in IDEs.
matplotlib-base	Creating 2D plots and visualizations in Python.
markdown	Parsing and converting Markdown files to other formats
nbformat	Reading and writing Jupyter notebook files
nltk	Natural language processing tasks in Python, such as tokenization, stemming, and part-of-speech tagging.
nomkl	Allows other packages to use multiple versions of numpy without conflicts.
numpy	Scientific computing with support for arrays, matrices, and mathematical functions.
openpyxl	Reading and writing Excel files in the OpenXML format
pandas	Data manipulation and analysis, providing data structures such as DataFrame and Series.
packaging	Provides core utilities for Python packages.
pillow	Opening and manipulating image files
plotly	Creating interactive visualizations in Python
py4j	Connecting to Java processes from Python
pyarrow	Working with columnar data and parquet files
python	Base package for the Python programming language
python-dateutil	Working with dates and times
pytz	Working with time zones
pyyaml	Reading and writing YAML files in Python
requests	Making HTTP requests in Python
scikit-learn	Machine learning tasks in Python, such as classification and regression.
scipy	Scientific computing in Python, with support for linear algebra, optimization, and signal processing.
seaborn	Creating statistical visualizations in Python
statsmodels	Statistical modeling in Python, with support for linear regression and other models.

For Windows OS, start Anaconda Prompt (Run as Administrator) and run the following code[5]:

```
conda create -n my_python_env python=3.9 knime-python-base
knime-extension geopandas osmnx keplergl pulp pandaral-
lel mgwr pysal spreg python-igraph leidenalg -c knime -c
conda-forge
```

This code creates a new environment `my_python_env` and installs the packages listed in Table 1.2 with Python 3.9.

TABLE 1.2

Python Packages Used in Geospatial Analytics Extension for KNIME

Package Name	Function
knime-python-base	Allows users to execute Python scripts within the KNIME Analytics Platform. It is used to integrate Python into the KNIME workflow.
knime-extension	Provides additional functionality to the KNIME Analytics Platform, such as new nodes, data types, and services.
geopandas	Extends the functionality of pandas to work with geospatial data. It allows users to perform operations on geometric types like points, lines, and polygons, in addition to the data analysis capabilities of pandas.
osmnx	Allows users to easily download, model, analyze, and visualize street networks from OpenStreetMap
keplergl	Allows users to create and interact with geospatial data visualizations in a web-based interface
pulp	Provides a linear programming solver for Python. It is used for solving optimization problems, such as finding the maximum or minimum of a function subject to constraints
pandarallel	Allows users to easily parallelize the execution of pandas operations on multiple cores.
mgwr	Allows users to fit geographically weighted regression models in Python.
pysal	Provides a set of tools for spatial data analysis, including functions for spatial weighting and spatial econometrics.
pygeoda	Perform spatial analysis functions of Geoda, a popular software for spatial data analysis.
spreg	Provides tools for estimating spatial regression models in Python, including spatial autoregressive models and error component models.
pointpats	Provides statistical analysis of planar point patterns

Go to File > Preference > KNIME to start KNIME AP. As shown in Figure 1.7, (1) click Conda, and use Browse to specify the installation path to Anaconda, (2) click Python, and check Conda under Python environment configuration. The installed Python 3 Conda environment is automatically detected and shown under Python 3 (default). Click Apply and Close.

Step 5. Nodes in Geospatial Analytics Extension for KNIME: In the Geospatial Analytics Extension for KNIME, nodes are the building blocks that allow users to perform various operations on geospatial data. These nodes can be connected to create a workflow that performs a series of tasks. Appendix 1 lists some of the tasks such as data import, cleaning, transformation, analysis, and visualization.

Step 6. Python and R script nodes: KNIME Python Integration has two nodes that allow users to execute Python scripts within the KNIME workflow. *Python Script* node allows users to write and execute Python code within the KNIME workflow. It can be used to perform various types of tasks, such as data cleaning, transformation, analysis, and visualization. The *Python View* node enables users to create and export visualizations based on Python scripts. It should be noted that the Python Script node can be configured to include an image output that can be used as a Python View node.

KNIME Interactive R Statistics Integration allows users to seamlessly integrate R statistics into the KNIME environment. It provides a user-friendly interface for accessing and executing R functions and scripts directly within KNIME, making it easier for data scientists and analysts to use R for advanced analytics and statistical modeling. The extension supports the import of data from various sources and enables the integration of R output into the KNIME workflow.

The KNIME Interactive R Statistics Integration extension includes several nodes that allow users to incorporate R into their data analysis workflows such as:

- R Snippet node enables users to run R scripts within the KNIME environment. R to Table, Add Table to R and R to R enable the connection between table and R workspace.
- R View node allows for visualization of R graphics and plots.
- R Learner node enables users to fit R machine learning models, and R Predictor node allows for prediction with R models created in the R Learner node.

These nodes are designed to be flexible and easy to use and provide a convenient way for users to harness the power of R for advanced analytics and statistical modeling without having to leave the KNIME environment.

1.3   CASE STUDY 1: MAPPING AND ANALYZING POPULATION DENSITY PATTERN IN BATON ROUGE, LOUISIANA

This section uses a case study to illustrate the implementation of some popular GIS tasks in KNIME Analytics Platform (AP). The study explores the urban population density in Baton Rouge, Louisiana. Users will learn how to implement attribute query, spatial query, attribute join, spatial join, projection, geometric attribute calculation, buffer, spatial overlay, dissolve, and intersect in KNIME AP.

Data needed for the case study are provided under the folder BatonRouge[6]:

1. Features County.zip and Tract.zip are the counties (referred to as "parishes" in Louisiana) and census tracts in Louisiana, respectively, both with a population field POP100,
2. Feature BRCenter.zip is the city center of Baton Rouge, and
3. Feature BRBLK.zip contains census blocks in East Baton Rouge Parish (EBRP) with a population field as Popu2020.

Figure 1.8 shows the workflow of Case Study 1. The workflow file Case1.knwf is provided under the folder for reference.

1.3.1   Creating Workflow, Reading and Visualizing Data

Step 1. Creating a workflow group and a workflow: Before starting the work, we need to create a new workflow under a workflow group, which can be seen as a file under a folder.

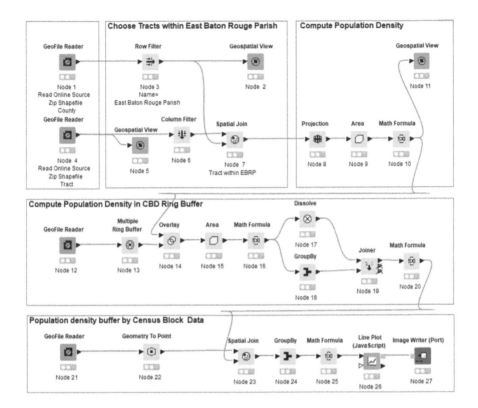

FIGURE 1.8   Workflow of Case Study 1

Create a workflow group. In KNIME Explorer (Figure 1.9a) > right click 🔧 LOCAL (Local Workspace). In the bring-up menu (Figure 1.9b) > 📑 New Workflow Group. In the dialogue of New KNIME Workflow Group Wizard, input CMGISV3 as Name of the workflow group to create, keep the default setting in Destination of new workflow group > click Finish (Figure 1.9c). A new work group CMGISV3 with a folder icon is created under LOCAL (Local Workspace). This folder will be used for the whole book. Repeat the previous step, create a new workflow group Chp01 under the folder 📁 CMGISV3 (Figure 1.9d).

Create a new workflow. Still in KNIME Explorer > right click the new folder Chp01 under CMGISV3 > ⬜ New KNIME Workflow. In the dialogue of New KNIME Workflow Wizard, input Case1 as Name of the workflow to create, keep the default setting in Destination of new workflow group > click Finish. A new workflow Case1 is created under the workgroup Chp01, and automatically opened in the Workflow Editor (Figure 1.9e).

(Optional) **Add a data folder as workflow group.** There are 2 ways to add another folder under the workgroup Chp01: (1) use the same step in Create workflow group to add a new one under the workgroup Chp01, and name it as data; or (2) go to the workgroup folder (e.g., E:\KNIME\Chp01, using

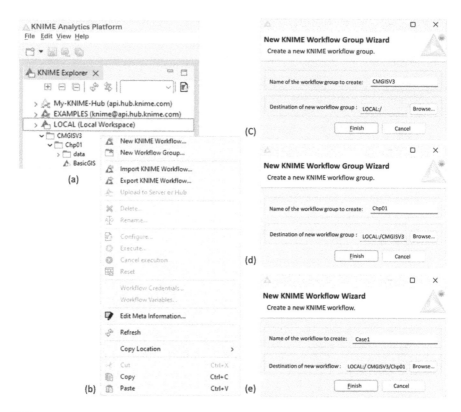

FIGURE 1.9 Interfaces: (a)-(d) Create New Workflow Group, and (e) Create Workflow

File Explorer in Windows), and create a new data folder. Then copy all the files into the data folder ☐. In KNIME Explorer > right click the workflow group Chp01 > 🔄 Refresh. All the new added data files are updated.

The following steps show how to build the workflow Case1 for this case study step by step.

Step 2. Reading the Shapefiles with GeoFile Reader: In Node Repository > 🎯 Community Nodes > 🔵 Geospatial Analytics > ⬛ Spatial IO > drag ⬛ GeoFile Reader to the Editor canvas. The GeoFile Reader node (Node 1) consists of two parts, as shown in Figure 1.10a, node icon and node status bar. Wherein, the black triangle on the right side shows that this node contains an output tabular data portal, and the traffic-light like bar indicates the contemporary node status among the four scenarios such as Not configured, Configured, Executed and Error.

In the Editor canvas, double click GeoFile Reader (Node 1) to bring up the pop-up dialog. As shown in Figure 1.10b, input the file path for the County boundary data in Louisiana, e.g., E:\KNIME\CMGISV3\Chp01\data\County.zip, or input a URL path,[7] and then click OK to confirm the setting and turn off the dialog. Right click the GeoFile Reader (Node 1) > choose Execute (Figure 1.10c).

After execution, the status bar turns to a green light. View the result table by right clicking the GeoFile Reader (Node 1) and choosing 🔲 Geodata table (Figure 1.10c).

FIGURE 1.10   GeoFile Reader node: (a)-(c) interfaces, and (d) output table

Figure 1.10d shows the result table, which contains the attributes of the County data, such as STATEFP, GEOID, NAME, . . . , and its geometry information represented by the column ⊕ geometry. The values in the geometry column indicate that most of the geometry units are polygons in a coordinate reference system (CRS) EPSG:4269. Right-click on geometry and choose Available Renders to change the rendering method of geometry information such as WKT, String, etc.[8] The tab Spec-Columns:17 provides the summary information about the columns, such as Column Type, Lower Bound, Upper Bound, etc. The tab Flow Variables displays the name and values for all flow variables passed by the workflow, which is helpful in workflow control and automation.

Double click the note "Node 1" under the status bar of GeoFile Reader to edit the node annotation and then provide concise explanations for other users or collaborators.

Step 3. Applying Visualization with Geospatial View node: To view the data on a map, we can use the nodes in the category 🎯 Spatial Visualization of ⊕ Geospatial Analytics. Drag 🌑 Geospatial View node to the Editor canvas. This node only has a black triangle on the left side, meaning that it only has an input portal for tabular data. Click the black triangle of GeoFile Reader (Node 1) and then hold the left mouse button to drag a link (line) to link it to the input portal of Geospatial View (Node 2). These two nodes and the link between them are considered a prototype workflow for reading and mapping spatial data. Hereafter in the book, the step is simply referred to as "link GeoFile Reader (Node 1) to Geospatial View (Node 2)".

Double click Geospatial View (Node 2) to bring out the dialog. It contains two components: preview (Figure 1.11a) and parameter settings (Figure 1.11b). On the parameter setting panel, (1) keep geometry for Geometry column, (2) uncheck Stroke, (3) under Marker tooltip columns, choose STATEFP and NAME in the Includes box by using ">" or double clicking the variable under Excludes, (4) similarly, under Marker popup columns, choose NAMELSAD, (5) under Size Settings, keep the default settings for Marker size column and Marker size scale, (6) under Coloring Settings, choose ALAND for Marker color column, Purples for Color map, keep Classify numerical maker color column as checked, choose FisherJenks for Classification method, 5 for Number of classes, (7) under Base Map Setting, keep the default OpenStreetMap for Base Map, (8) keep the default setting under Legend Setting, then (9) click the button Save & execute to visualize the geospatial data.

As shown in Figure 1.11c, the values in the column ALAND (land area) are classified as five categories based on the classification method Fisher Jenkens and

FIGURE 1.11   Interface of Geospatial View node:(a) preview window, (b) setting pane, and (c) result

visualized in the color of Purples. The Open Street Map is used as a base map. The values of STATEFP and NAME of polygons change with the mouse hovering on them, and the value of NAMELSAD of a county appears while it is clicked.

Users can change the parameter settings and click Save & execute to preview it. When satisfied with the result, click OK to accept the setting and turn off the dialog. In the Editor canvas, right click Geospatial View (Node 2) and choose Execute. After execution, right click the node and choose Interactive View: Geospatial View to view the map, on which we can zoom, pan and click.

1.3.2 Extracting the Study Area

Step 4. Extracting East Baton Rouge Parish boundary by Row Filter: In Node Repository> ✦ Manipulation > ▬ Row > ⬚ Filter > drag ⇸ Row Filter to the Editor. Link GeoFile Reader (Node 1) to Row Filter (Node 3), and double click Row Filter (Node 3) to bring out the dialog. In the dialog (Figure 1.12), under Filter Criteria tab, check Include rows by attribute value on the left list; under Column value matching, choose NAMELSAD in the drop-down menu for Colum to test; under Matching criteria, check use pattern matching, input East Baton Rouge Parish in the box; and click OK. Execute Node 3.

After execution, right-click Row Filter (Node 3) and choose ▤ Filtered to view the result table. It shows one row. Users can link Row Filter (Node 3) to Geospatial View (Node 2) and re-execute Geospatial View (Node 2) to view the East Baton

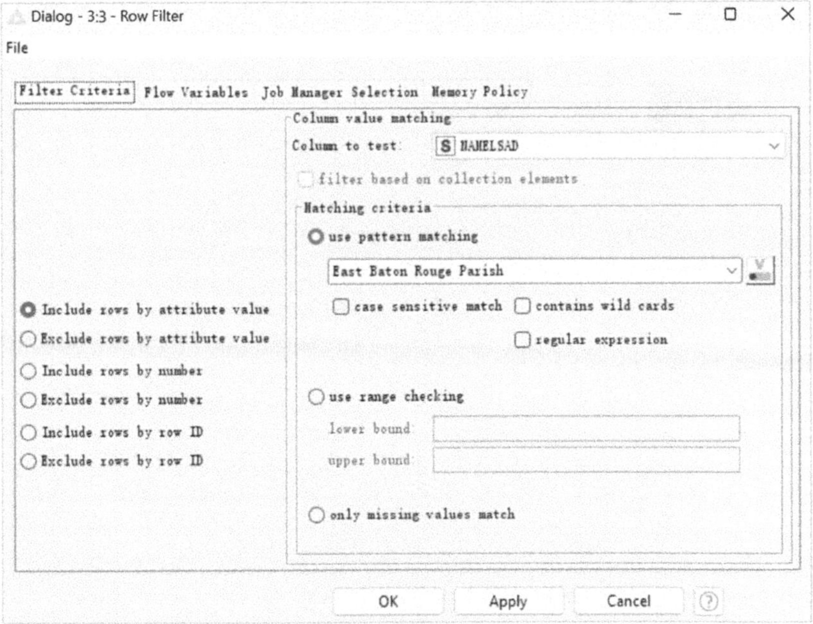

FIGURE 1.12 Dialog of Row Filter

Rouge Parish (EBRP) boundary on the map remember to uncheck `Classify numerical maker color column`, as there is only one row.

Step 5. Reading census tract data and filter columns: Drag another GeoFile Reader node (Node 4) to the Editor canvas and repeat Step 2 to read the census tract data by setting the file path as `E:\KNIME\CMGISV3\Chp01\data\Tract.zip`, or an online URL address.[9]

To visualize the data, we can add another Geospatial View node (Node 5) and link GeoFile Reader (Node 4) to it. In the dialog, choose `POP100` (or other fields) in the Marker color column to view the corresponding map.

In order to speed up the data processing and reduce the amount of data involved, it is recommended to filter out unnecessary columns and keep only the key variables required for a particular task. To achieve this, here we can add a Column Filter node (Node 6) and connect it to the GeoFile Reader (Node 4). In the dialog box shown in Figure 1.13, select the key columns `GEOID`, `POP100` and `geometry` by double-clicking on their names in the Includes box, or by selecting them and using the arrow button to move them to the other side. Then, click OK and execute the node.

Step 6. Applying spatial join to extract census tracts in EBRP: This step uses the EBPR boundary generated by Row Filter (Node 3) to extract the census tracts within it.

In ⊘ Geospatial Analytics > ⊙ Spatial Manipulation, drag ⊙ Spatial Join to the Editor canvas. Link Column Filter (Node 6) and Row Filter (Node 3) to the two input ports of Spatial Join (Node 7), and double click Spatial Join (Node 7) to bring out the dialog. As shown in Figure 1.14a, keep `geometry` for both Left geometry

FIGURE 1.13 Dialog of Column Filter

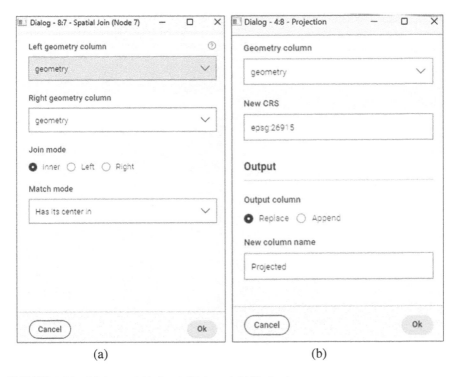

FIGURE 1.14 Dialogs of (a) Spatial Join and (b) Projection

column and Right geometry column, check `Inner` for Join mode, choose `Has its center in` for Match mode, click OK and execute the node. All the census tracts with their representative points falling within the boundary of EBRP are retained and associated with the attributes of EBRP.

Both the nodes Spatial Join and Overlay use the Join mode to specify what type of join occurs and which geometry is retained in the result. In addition to `Inner`, it also accepts the options such as: `Left` to retain all rows from the left and only match rows from the right input tables, and `Right` to retain all rows from the right and only match rows from the left input tables (Figure 1.15a). The match mode specifies how to join the attributes of one object to another based on their geometric relationship (Figure 1.15b).[10]

Please note that when using the Spatial Join or Overlay node, duplicate column names will be automatically appended with the suffix "_1" and "_2". This ensures that the column names remain unique and distinct within the resulting overlayed dataset.

Step 7. Calculating population density: This step utilizes three nodes to transform the coordinates reference system (CRS) and calculate the area and population.

Under ⊛ Geospatial Analytics > ⬡ Spatial Transformation, drag ⊞ Projection to the Editor canvas. Link Spatial Joiner (Node 7) to Projection (Node 8). In the dialog of Projection (Node 8) shown in Figure 1.14b, keep `geometry` for Geometry column, input `epsg:26915` for New CRS, check `Replace` under Output column, and click OK to execute the node. This node transforms the input

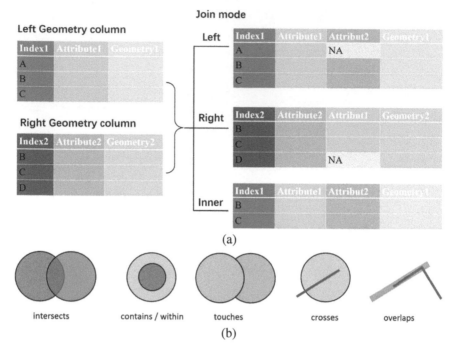

FIGURE 1.15   (a) Join mode and (b) Match mode for spatial relationship

data in a geographic CRS in decimal degrees (egsg:4326) to a projected CRS (epsg:26915).[11]

Under ⊞ Spatial Calculation, drag ⬭ Area to the Editor, link Projection (Node 8) to Area (Node 9), and execute the node. It calculates the area based on the CRS unit of input data, and attaches a new column named `area` to store the geometric value in meter.

Under ⚲ Manipulation > ▮ Column > ▢ Convert & Replace), drag ꬍ Math Formula to the Editor canvas, and link Area (Node 9) to Math Formula (Node 10). In the dialog shown in Figure 1.16, (1) click the Math Expression tab, input `$POP100$*1000000/ $area$` in the box under Expression by double clicking the variable names (`POP100`, `area`) under Column List and the operation symbol (/) under Function; (2) check Append Column, and input `PopDenSqkm` in the box, and (3) click OK and execute the node. This node appends a new column `PopDenSqkm` for population density of each census tract.

Add a Geospatial View node (Node 11) and link Math Formula (Node 10) to it. Configure it to visualize the field `PopDenSqkm`.

1.3.3 Examining Urban Population Density Pattern Based on Census Tract Data

This subsection generates multiple ring buffers around the city center and examines how the population density declines with distance from the city center based on data at the census tract level.

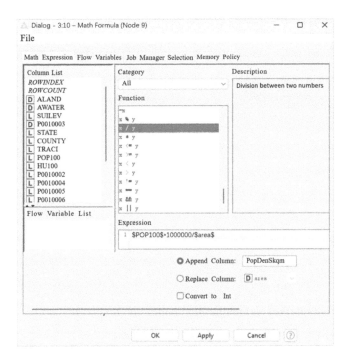

FIGURE 1.16 Interface of Math Formula

Step 8. Creating Multiple Ring Buffers: Under ⬡ Geospatial Analytics > ▦ Spatial IO, drag ▦ GeoFile Reader to the Editor canvas, and input the file path as E:\KNIME\CMGISV3\Chp01\data\BRcenter.zip, or an online URL address.[12] Click OK and execute the GeoFile Reader (Node 12).

Under ⬡ Geospatial Analytics > ⬣ Spatial Manipulation, drag ⊚ Multiple Ring Buffer to the Editor canvas, and link GeoFile Reader (Node 12) to Multiple Ring Buffer (Node 13). In the dialog shown in Figure 1.17a, keep geometry for Geometry column, input 2,4,6, . . . ,26 in Serial buffer distances, choose Kilometer for Distance unit, check Keep CRS from input table, and click OK and execute the node.

The resulting table contains two columns, Distance and Buffer.

Step 9. Overlaying census tracts and ring buffers: Under ⬡ Geospatial Analytics > ⬣ Spatial Manipulation, drag ⬩ Overlay to the Editor canvas. Link Math Formula (Node 10) to the top input port of Overlay (Node 14), and link Multiple Ring Buffer (Node 13) to the bottom input port. In the dialog shown in Figure 1.17b, keep the default name geometry for Left geometry column and Buffer for Right geometry column, choose Intersection for Overlay mode, check Return only geometries of the same geometry type,[13] click OK and execute the node.

To view the result, link Overlay (Node 14) to Geospatial View (Node 5).

The overlay function contains tools to overlay multiple feature classes. All overlay operations involve joining two sets of features into a single set of features to identify

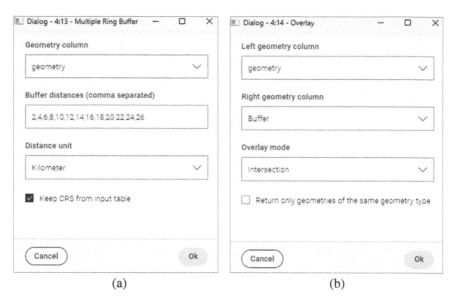

FIGURE 1.17   Dialogs of (a) Multiple Ring Buffer and (b) Overlay

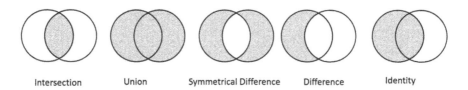

FIGURE 1.18   Examples for Overlay

spatial relationships between the input features (Figure 1.18). In addition to *intersection* that preserves only the overlaps of all input features, overlay contains other options for identifying spatial relationships: *union* (union of the input features), *symmetrical difference* (portions in the input features that do not overlap), *difference* (only those portions of the input features falling outside the erase features), and *identity* (the portions thereof that overlap identity features will get the attributes of those identity features).

Step 10. Calculating population density for each ring buffer: Under ⊕ Spatial Calculation, drag ▢ Area to the Editor, and link Overlay (Node 14) to Area (Node 15). Execute the node to calculate areas. Right click on the node and choose Geo table with area to view the result. A column `area` (#1) indicates area size for each intersected unit in square meters. In Geospatial Analytics Extension for KNIME, it automatically creates alias names for duplicated column names, as the default column name for area calculation is `area`.

Similar to Step 8, drag Math Formula to the Editor canvas, link Area (Node 15) to Math Formula (Node 16). In the dialog, (1) input `$PopDenSqkm$*($area(#1)$/1000000)` under Expression, and (2) check Append Column, and input `Popu` in the box, (3) check `Convert to Int`. Click OK to execute the node. This node calculates the total population for each intersected unit.

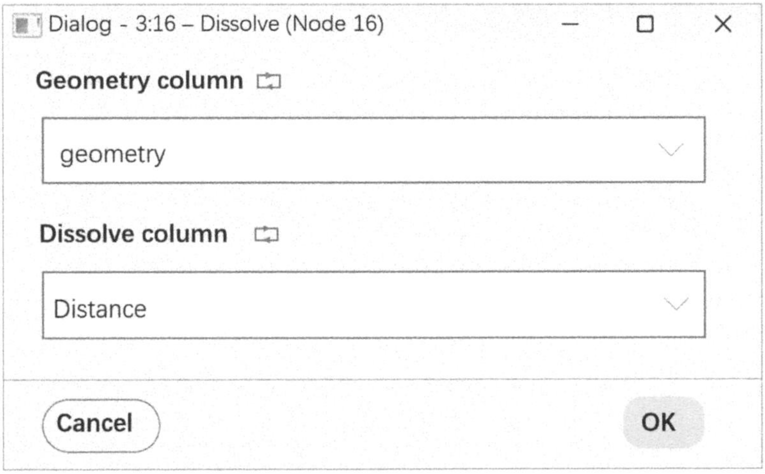

FIGURE 1.19   Interface of Dissolve

Under ⊕ Geospatial Analytics > ⊙ Spatial Manipulation, drag ⊗ Dissolve to the Editor canvas, and link Math Formula (Node 16) to Dissolve (Node 17). In the dialog (Figure 1.19), keep geometry for Geometry column, choose Distance for Dissolve column, and click OK to execute the node. The node aggregates the geometric information in ring buffers.

In Node Repository > ⚡Manipulation > ▬ Row > ▢ Transform, drag ⤵ GroupBy to the Editor, and link Math Formula (Node 16) to GroupBy (Node 18). In the dialog, under the default tab Settings and its sub-tab Groups (Figure 1.20a), (1) in the box of Group Setting, choose Distance for Group column(s); (2) under Advance setting, choose Keep original name(s) for Column naming and keep other default settings; (3) then click sub-tab Manual Aggregation (Figure 1.20b), double click (or use add) area(#1) and Popu in the Available column(s) to set them as input variables under Column in the right side box; (4) right click (or click) the cell under Aggregation (click to change) to choose Sum as aggregation method and (5) click OK and execute the node. This node calculates the total area and population for each ring buffer.

Step 11. Joining attributes of ring buffer: In this step, the aggregated geometry generated by the Dissolve node and the summarized attributes produced by the GroupBy node for the ring buffers are joined using an attribute join performed by the Joiner node.

In Node Repository> ⚡Manipulation > ▮ Column > ▢ Split & Combine, drag ⥃ Joiner to the Editor canvas. Link Dissolve (Node 17) to the top (or called left) input port of Joiner (Node 19), and link GroupBy (Node 18) to the bottom (or called right) port of Joiner (Node 19). Double click Joiner (Node 19) to bring out the dialog.

Click the Joiner Settings tab (Figure 1.21a), (1) under Join Columns, check all of the following for Match, click Add matching criterion, then choose Distance for Both Top Input ('Left' table) and Bottom Input ('right' table), check value and type for Compare values in join columns by, (2) check Matching rows under Include in output, and (3) keep other default settings in Output options and Row keys.

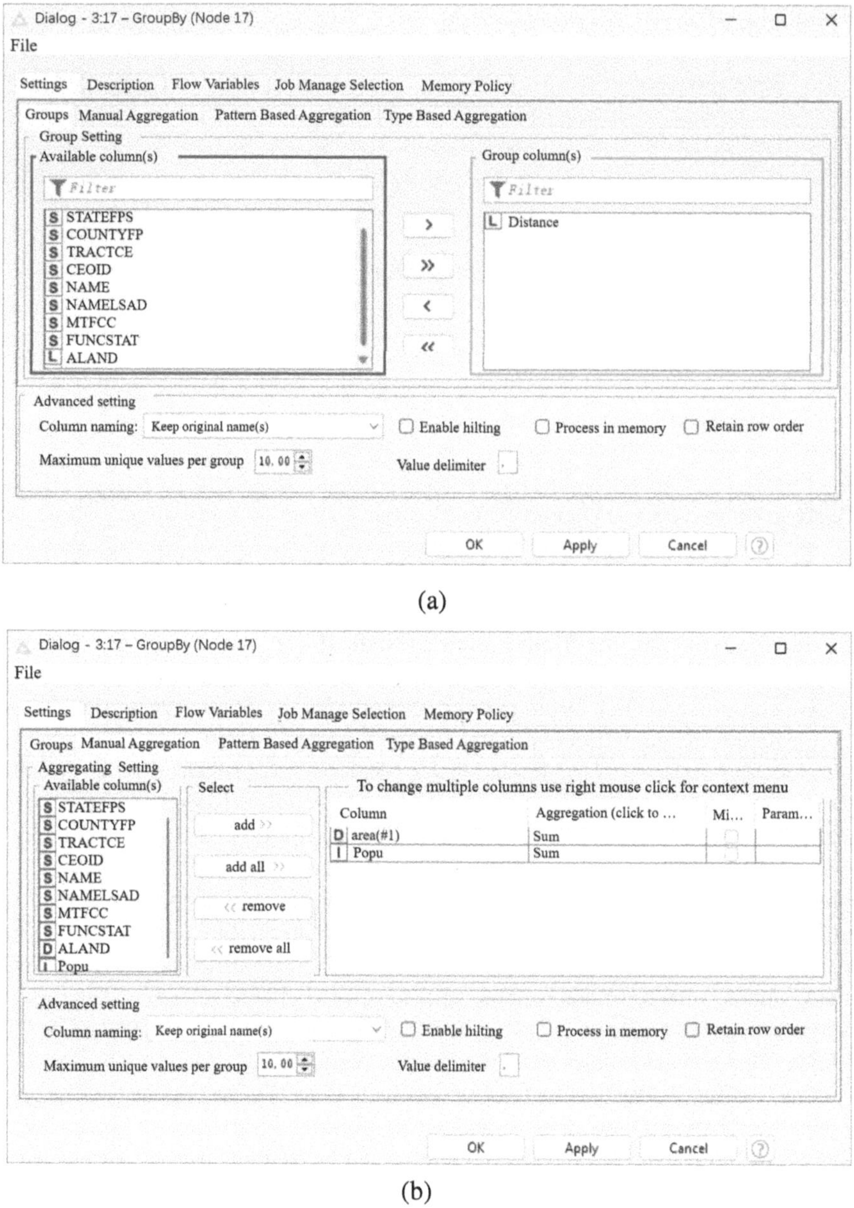

(a)

(b)

FIGURE 1.20   Interface of GroupBy: (a) Groups tab and (b) Manual Aggregation tab

Please note that when it comes to the option Compare values in join column by, the `string representation` will be used when joining columns with data types of Integer and String. Additionally, making integer types compatible will be applied for those with data types of Integer and Long.

Then click Column selection tab (Figure 1.21b), (1) under Top Input (left table), check `Manual Selection`, use < to only include `geometry` from the Top (left)

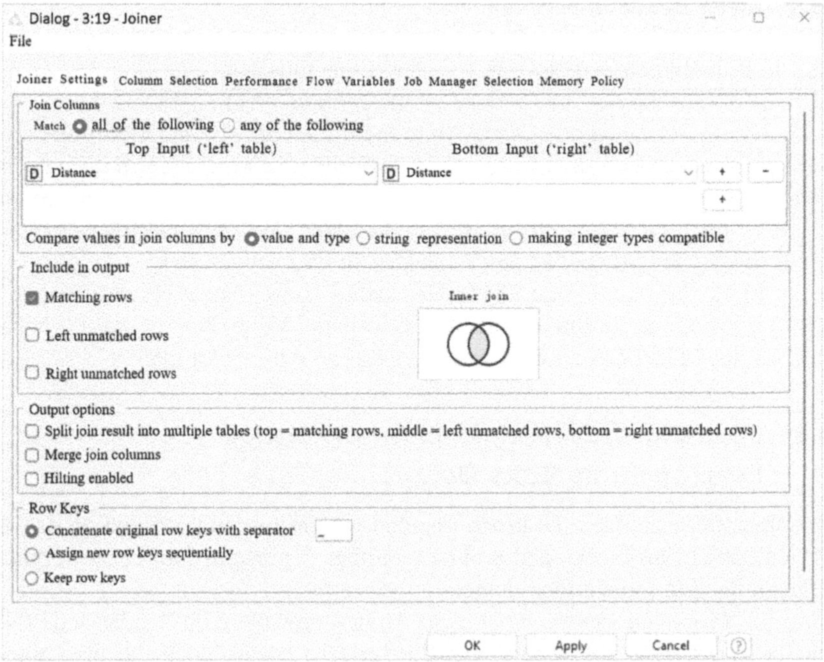

(a)

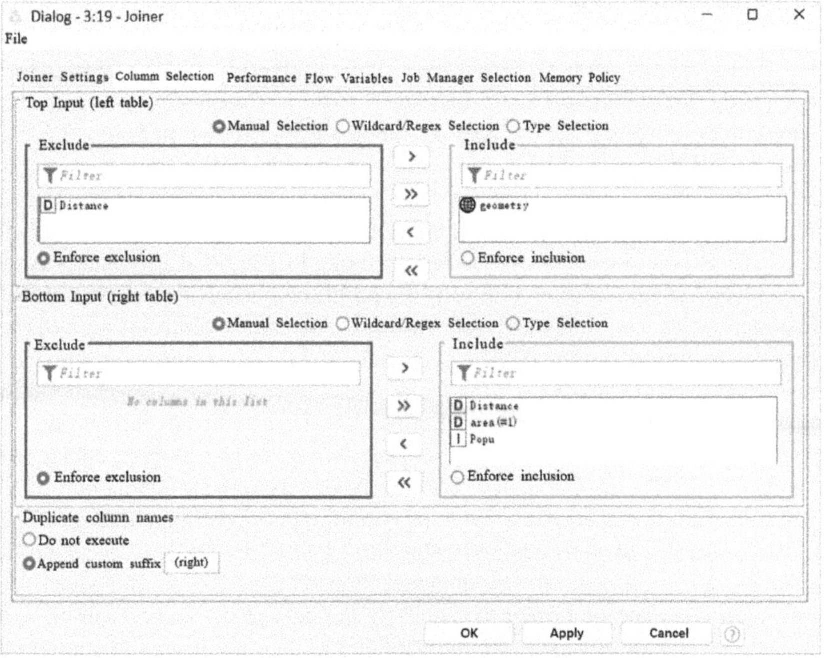

(b)

FIGURE 1.21   Interface of Joiner: (a) Join settings and (b) Column selection

data; (2) under Bottom Input (right table), check `Manual Selection`, use > to include all columns from the Bottom (right) data; (3) keep default settings under Duplicate column names. Click OK to execute the node.

This step joins the two tables based on the common field `Distance`.

Add a Math Formula node (Node 20) to the Editor canvas, and link Joiner (Node 19) to it. In the dialog, (1) input `$Popu$*1000000/ $area(#1)$` in the box under Expression, (2) check Append Column, and input `PopuDen` in the box, and (3) check `Convert to Int`. Click OK to execute the node. This node calculates the population density in persons per square kilometer for each ring buffer.

Link Math Formula (Node 20) to Geospatial View (Node 11) to visualize the result by revising the parameter setting in Geospatial View (Node 11), e.g., choosing `PopuDen` for Maker color column and `viridis_r` for Color map.

1.3.4 EXAMINING URBAN POPULATION DENSITY PATTERN BASED ON CENSUS BLOCK DATA

This subsection examines the urban population density pattern based on data at a finer scale in census blocks and provides another opportunity for users to practice similar skills with some minor changes.

Step 12. Preparing census block data: Drag a new GeoFile Reader node (Node 21) to the Editor canvas, and repeat Step 2 to read the census block data by setting the file path as `E:\KNIME\CMGISV3\Chp01\data\BRblk.zip`, or an online URL address.[14] Execute the node.

In Node Repository > 🐝 Community Nodes > 🌐 Geospatial Analytics > 🌐 Spatial Transformation, drag 🔵 Geometry to Point to the Editor canvas. Link GeoFile Reader (Node 21) to the new node Geometry to Point (Node 22), and double click Geometry to Point (Node 22) to bring out the dialog. As shown in Figure 1.22, keep `geometry` in Geometry column, check `centroid` under Point Type Selection, check `Append` for Output Column, keep the default name `Point` for New Column Name, and click OK and execute the node. This step adds a new geometry column called `Point` to the existing data. Having a table with multiple geospatial attributes is a unique feature of KNIME, unlike other GIS software programs such as ArcGIS Pro or QGIS.

This step yields a point feature representing census block centroids with the same coordinates reference system as the ring buffer. As block is a finer unit than tract, it is reasonable to aggregate population from blocks to ring buffers by assuming that population in a ring buffer is composed of blocks whose centroids falls inside.

Step 13. Aggregating population from census blocks to ring buffers: Add a Spatial Join node (Node 23), link Geometry To Point (Node 22) to the top input port of Node 23, and link Math Formula (Node 20) to its bottom input port. In the dialog of Spatial Join, choose `Point` for Left geometry column, `geometry` for Right geometry column, keep `Inner` checked under Join mode, choose `Intersects` for Match mode, and click OK and execute it (refer to step 7). In the data of output port, all the census block centroid data are attached with the attributes of their corresponding buffer rings.

Add a GroupBy node (Node 24) and link Spatial Join (Node 23) to it. In the dialog of GroupBy (Node 24), under the default tab Settings and its sub-tab Groups, (1) in the box of Group Setting, choose `Distance` for Group column(s); (2) under

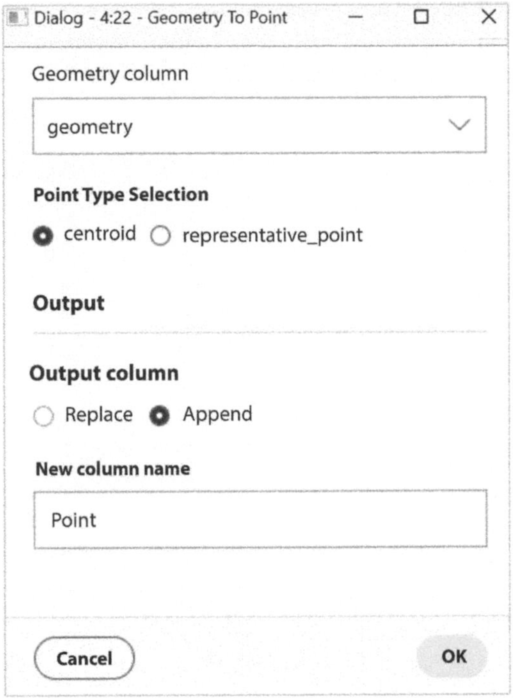

FIGURE 1.22   Dialog of Geometry to Point

Advance setting, choose `Keep original name(s)` for Column naming and keep other default settings; then click sub-tab Manual Aggregation, (3) move `Area`, `Popu2020` and `PopuDen` in the Available column(s) to Column in the right side box, then set the aggregation methods as `Sum`, `Sum`, and `Mean`,[15] respectively. Click OK to execute the node.

The node calculates the total area and population for all census block centroids that fall in the same ring buffer, and keep the original population density value `PopuDen` derived from the census tract data (refer to step 11).

Add a Math Formula node (Node 25) to the Editor canvas, and link GroupBy (Node 24) to it. In the dialog of Math Formula (Node 25), (1) input `$Popu2020$*1000000/$Area$` in the box under Expression, (2) check Append Column, and input `PopuDenBlk` in the box, and (3) check `Convert to Int`. Click OK to execute the node. This calibrates the population density in each ring buffer in the unit of persons per square kilometer, based on data in census blocks.

Step 14. Plotting the population density pattern and saving the plot result: Under ⌕ Views > ᴊꜱ JavaScript, drag ⟋ Line Plot to the Editor Canvas, and link Math Formula (Node 25) to Line Plot (Node 26). It yields two output ports: input data (bottom) and image (green block on the top).

In the dialog, (1) under the Option tab (Figure 1.23a), check `Create image at outport`, choose `Distance` for Choose column for x-axis, (2) under Choose column for y-axis, double click the variable name in the Exclude box (or choose the variable

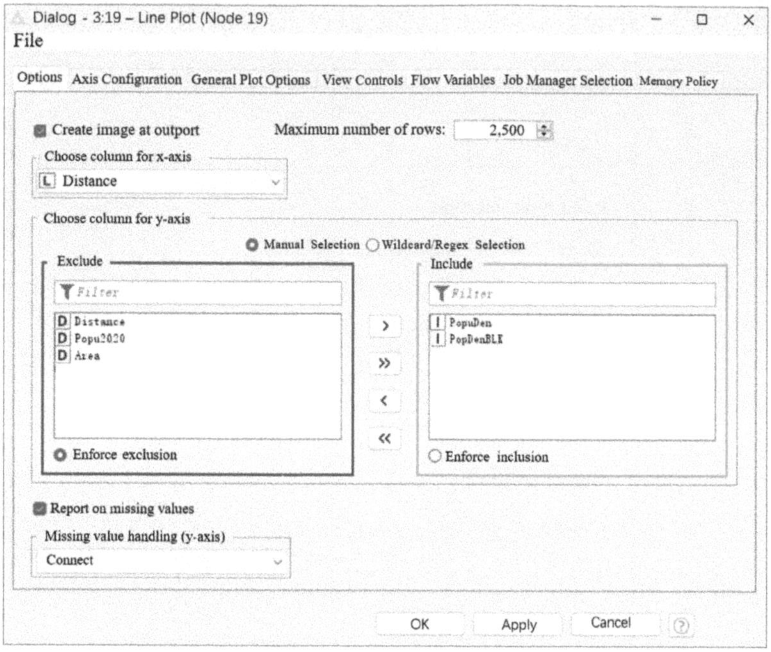

(a)

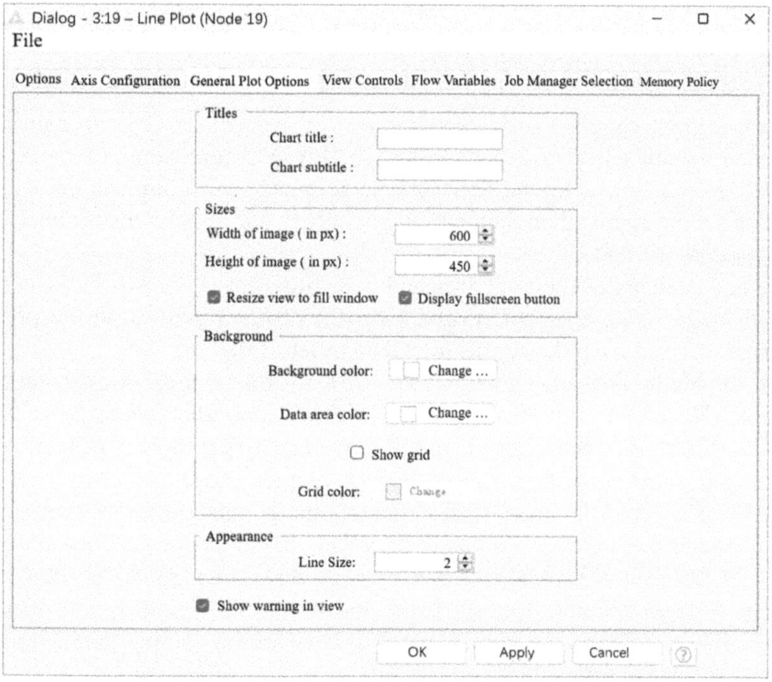

(b)

FIGURE 1.23   Interface of Line Plot: (a) Options and (b) General Plot Options

and click arrow sign such as $\rangle$) to set PopuDen and PopuDenBlk in the Include box, and keep other default setting; (3) click General Plot Options tab (Figure 1.23b), input names for Titles, e.g., Population Density Pattern, input 600 for Width of image and 450 for Height of image,[16] and (4) uncheck Show grid under Background, and keep other default settings. Click OK to execute the node.

Right click Line Plot (Node 19), choose Q Interactive View: Line Plot to view the plot, or choose Image to view the output image (Figure 1.24).

Then we will use Image Writer (Port) node to save it to a local disk. In Node Repository > ⛏ IO > ⊡ Write, drag ⊡ Image Writer (Port) to the Editor, and link Line Plot (Node 26) to it. In the dialog of Image Writer (Port) (Node 27) shown in Figure 1.25, under the Settings tab, (1) choose Local File System for Write

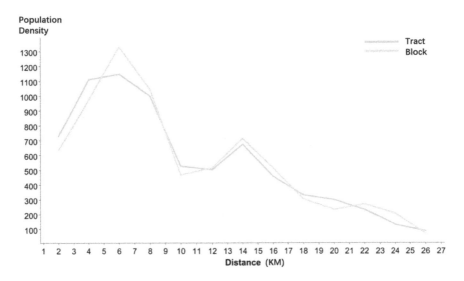

FIGURE 1.24 Population density pattern across ring buffers: Tracts vs. Blocks

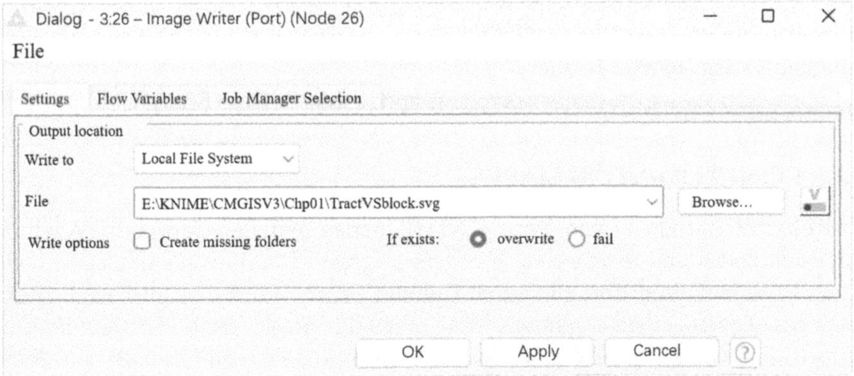

FIGURE 1.25 Dialog of Image Writer (Port)

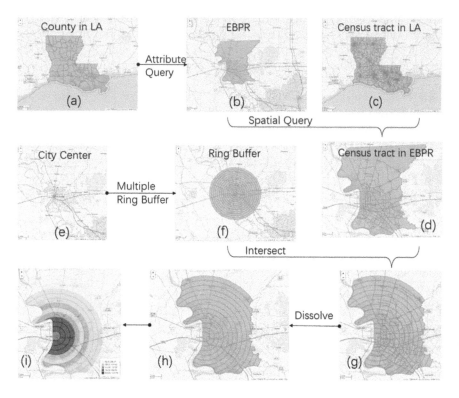

FIGURE 1.26 Major operations and results of Case Study 1

to, (2) Browse (or input) the file path to save the figure, e.g., E:\KNIME\CMGISV3\ Chp01\TractVSblock.svg, and (3) check overwrite for If exists. Click OK to execute it. The image is saved in the designated path.

It can be also saved with a relative path by choosing Relative to and Current workflow for Write to, and input ../ TractVSblock.png for File. It will be saved with the relative path in the workflow group Chp01. To save the table with the geometry column, use the node GeoFile Writer in Spatial IO, which shares the same parameters with GeoFile Reader.

Figure 1.26 shows the major operations and results of Case Study 1.

1.4 CONCLUDING REMARKS

This chapter introduces the basic KNIME nodes and the Geospatial Analytics Extension nodes and their usage. It covers a range of built-in manipulation nodes in KNIME, including Row Filter and Column Filter, which can filter rows or columns based on specified conditions using manual or Regex methods. Math Formula performs mathematical operations on columns in a table, and GroupBy and Joiner nodes are popular nodes that group rows by specified columns and perform aggregate functions on the resulting groups. Joiner is used to join tables based on common fields. Additionally, visualization nodes like Line Plot can be used to explore data, and Image Writer node can save the plot. These basic KNIME nodes are useful for manipulating, visualizing, and analyzing data.

The chapter also covers tools in the Geospatial Analytics Extension for KNIME. Spatial IO nodes like GeoFile Reader read Shapefile or GeoJSON data. Spatial Visualization nodes like Geospatial View visualize geospatial data. For Spatial Transformation nodes, Projection converts geospatial data between different coordinate reference systems, and Geometry to Point creates centroids or representative points from polygons or lines.

In addition, several Spatial Manipulation nodes are introduced. Spatial Join node can be used as spatial queries to join geospatial data based on spatial relationships. Overlay node is used to create a new feature based on the input features. Dissolve node dissolves polygons based on a specified attribute field. Additionally, Area node under the Spatial Calculation category calculates polygon area sizes in a geospatial dataset.

Overall, this chapter is designed for beginners who want to use KNIME nodes to process data and perform analysis. These basic KNIME nodes enable users to manipulate, visualize and analyze data. The Geospatial Analytics Extension nodes provide additional functionality for working with geospatial data and performing more complex spatial analysis tasks.

APPENDIX 1 NODE LIST IN GEOSPATIAL ANALYTICS EXTENSION FOR KNIME

Table 1.3 enumerates the nodes available in the Geospatial Analytics Extension for KNIME for tasks including data import, cleaning, transformation, analysis, and visualization. As this extension is continually being developed and updated, users can visit GitHub or KNIME Hub to access the latest nodes.

TABLE 1.3

Node List in Geospatial Analytics Extension for KNIME

Category	Node Name	Short Description
Spatial IO	GeoFile Reader	Read ESRI shapefile and GeoJSON
	GeoFile Writer	Write ESRI shapefile and GeoJSON
	GeoPackage Reader	Read GeoPackage layer (GPKG, GDB)
	GeoPackage Writer	Write GeoPackage layer
Spatial Calculation	Area	Calculates the area of geometric objects
	Length	Calculates the length of geometric objects
	Coordinates XYZ	Extracts the XYZ coordinates
	Bounds	Computes the bound for geometric objects
	Total Bounds	Computes the total bounds for all given geometric objects

(Continued)

TABLE 1.3

Continued

Category	Node Name	Short Description
	Bounding Box	Computes a minimum rectangular boundary that encloses a set of geographic features
	Bounding Circle	Computes a minimum circular boundary that encloses a set of geographic features
	Convex Hull	Computes a polygon that encloses a set of geographic features while minimizing its area and maintaining its convexity.
	Unary Union	Merges multiple polygon features into a single feature.
Spatial Manipulation	SpatialJoin	Combines the attribute data of two spatial datasets based on their spatial relationship.
	NearestJoin	Combines the attribute data of two spatial datasets based on their distance relationship.
	Overlay	Combines two or more spatial datasets into a single output dataset based on their spatial relationship
	Dissolve	Combines adjacent polygons with the same attribute value into a single polygon.
	Clip	Extracts a portion of a spatial dataset that intersects with a given boundary.
	Simplify	reduces the complexity of a spatial dataset while preserving its essential shape and spatial relationships.
	Buffer	Creates a polygon that surrounds a geographic feature at a specified distance.
	Multiple Ring Buffer	Creates multiple concentric buffers around a geographic feature at specified distances.
	Euclidean Distance	Calculates the distance between two points in a straight line.
	Haversine Distance	Calculates the distance between two points on a sphere using the Haversine formula.
	Create Grid	Creates a grid of equally sized polygons over a spatial extent.
	Voronoi (Thiessen) polygons	Partitions a geographic space into contiguous polygons based on the proximity to a set of point locations, where each polygon is defined as the area that is closest to one point relative to all other points in the dataset.
Spatial Transformation	Projection	Project spatial data from one coordinate system to another
	Geometry To Point	Extracts the centroid or a representative point from each feature in a geometry dataset

Category	Node Name	Short Description
	Polygon To Line	Converts polygon features into line features by creating the boundary lines of each polygon.
	Points To Line	Connects point features in the order they appear in a dataset, creating a single continuous line feature.
	Line To MultiPoint	Converts a line feature into a set of point features, where each point represents a vertex of the original line.
	Multipart To Singlepart	Separates multi-part features into individual single-part features.
	Create Random Points	Generates a set of random point features within a specified extent or on the surface of existing features.
	Line Endpoints	Calculates two geometry columns representing the start and end points of each line in the input data
Spatial Conversion	Lat/Lon to Geometry	Converts the given latitude and longitude column to a geometric point column
	GeoJSON to Geometry	Converts the input GeoJSON column to a geometry column
	WKT to Geometry	Converts the input Well-known-text (WKT) column to a geometry column
	Geometry to Lat/ Long	Extracts the latitude and longitude
	Geometry to GeoJSON	Converts the input geometry column to a Well-known-binary (WKB) column
	Geometry to WKT	Converts the input geometry column to a Well-known-text (WKT) column
	Geocoding	Converts a textual address or location information into geographic coordinates (latitude and longitude)
	Reverse Geocoding	Converts geographic coordinates (latitude and longitude) into a textual address or location information
	Geometry to Metadata	Extracts metadata information (e.g., CRS, type for each geometry) from geometric features
Spatial Visualization	Geospatial View	Displays geographic data on an interactive map, allowing users to pan, zoom, and explore the data in real-time.
	Geospatial View Static	Creates a static map visualization of geographic data with more options for image exportation
	Keplergl Geoview	Provides a powerful and customizable platform for visualizing large-scale geospatial data, featuring advanced mapping, filtering, and 3D visualization.
	Spatial Heatmap	Visualizes the density of point features in a geographic dataset by creating a heat map that highlights areas with a high concentration of points.

(Continued)

TABLE 1.3

Continued

Category	Node Name	Short Description
Exploratory Spatial Data Analysis	Spatial Weights	Defines the spatial relationships between features in a dataset, often used for spatial analysis or modeling
	Global Geary's C	Measures the spatial autocorrelation of a dataset, quantifying the similarity or dissimilarity of feature values among neighboring features
	Global Getis-Ord G	
	Global Moran's I	
	Local Getis-Ord G	Measures the degree of spatial clustering of high or low feature values in a dataset, identifying areas with significant hotspots or coldspots.
	Local Moran's I	
Spatial Modelling	OLS with Spatial Test	Ordinary least squares with results and diagnostics.
	2SLS with Spatial Test	Spatial two stage least squares (S2SLS) with results and diagnostics
	Spatial Error Model	Spatial Error Model by Maximum Likelihood
	Spatial Lag Model	Spatial Lag Model by Maximum Likelihood
	Spatial Lag Panel Model	Spatial Lag Panel Model with Fixed Effects
	Spatial Error Panel Model	Spatial Error Panel Model with Fixed Effects
	GWR Model	Geographically Weighted Regression
	GWR Predictor	Geographically Weighted Regression Predictor
	MGWR Model	Multiscale Geographically Weighted Regression
Location Analysis	P-median Solver	Identifies the optimal location for a set of facilities that minimizes the total travel distance or time between the facilities and their clients, subject to a specified number of facilities and other constraints.
	LSCP Solver	Solves the location set covering problem (LSCP), which involves identifying the minimum number of facilities needed to cover a set of clients, subject to various constraints such as capacity or distance limitations.
	MCLP Solver	Solves the maximum covering location problem (MCLP), which involves identifying the optimal location for a set of facilities that maximizes the number of covered clients, subject to various constraints such as capacity or distance limitations.
	P-center Solver	Identifies the optimal location for a facility that minimizes the maximum travel distance or time between the facility and its clients, subject to a specified number of facilities and other constraints.

Category	Node Name	Short Description
	MAEP Solver	Minimize inequality in accessibility of facilities and is currently formulated as minimal variance across geographic areas.
Spatial Clustering	SKATER	Implement the Spatial C(K)luster Analysis by Tree Edge Removal (SKATER) algorithm, providing an optimized algorithm to prune to tree into several clusters that their values of selected variables are as similar as possible.
	SCHC	Performs hierarchical clustering based on contiguity constraints, which groups features based on their common borders and maintains their spatial contiguity.
	REDCAP	Implement the REDCAP algorithm (Regionalization with dynamically constrained agglomerative clustering and partitioning)
	MaxP-Greedy	Considers the regionalization problem as an application of integer programming, the number of regions is determined endogenously.
	AZP-Greedy	The automatic zoning procedure (AZP) is a heuristic method for combining contiguous spatial units into p regions, with the aim of minimizing within-region variance
	Peano Curve	Assign spatial orders to points in 2D space based on space-filling curves
	MSSC Initialization	Performs modified scale-space clustering(MSSC) which groups features based on their spatial order along a Peano curve.
	MSSC Refiner	Refines the clustering using the MSSC algorithm, which iteratively adjusts the clustering to satisfy the criteria of ascending spatial order and minimum aggregation volume constraints.
	Isolation Tackler	Identifies and removes isolated features or clusters from the clustering results.
	Mean Center	Calculates the mean location of a set of points with its standard deviation distance
	Standard Deviation Ellipse	Calculates the ellipse that represents the dispersion or spread of the points within a cluster.
Spatial Network	Google Distance Matrix	Provides distance and travel time estimates between multiple origins and destinations based on different modes of transportation with the support of API key
	OSRM Distance matrix	Computes a distance, travel time or routes using the Open Source Routing Machine (OSRM) project
	Road Network Distance Matrix	Computes the shortest path distance or travel time between multiple points using a road network as a reference.

(Continued)

TABLE 1.3
Continued

Category	Node Name	Short Description
	Road Network Isochrone Map	Shows the areas that can be reached from a specific point within a certain amount of travel time or distance based on a road network.
Open Datasets	OSM Boundary Map	Get Boundary from the Open Street Map with Geocoding
	OSM Road Network	Get Road Network from the Open Street Map
	OSM POIs	Get Points of Interests (POIs) from the Open Street Map
	US2020 Census Data	Retrieve US 2020 Census Redistricting Data by state, county, tract, block group and block
	US2020 TIGER Map	Retrieve geospatial data from US Census TIGER/ Line 2010/2020 by state, county, tract, block group and block
	US ACS 5-Year Estimates	Retrieve American Community Survey 5-Year Data (2009–2020) of United States

APPENDIX 2   NODES SUMMARY

Table 1.4 provides a comprehensive overview of all the nodes utilized in this KNIME Lab Manual, with a total of 775 nodes used, based on 146 unique nodes (including components or different usages of one node). Among these, the top 10 frequently

TABLE 1.4
Node Summary

Nodes	Debut Appearance			Category[17]	Frequency
	Figure	Case Study	Section		
GeoFile Reader	1.10	1	1.3	**Spatial IO**	53
Geospatial View	1.11	1	1.3	**Spatial Visualization**	25
Row Filter	1.12	1	1.3	Manipulation	16
Column Filter	1.13	1	1.3	Manipulation	18
Spatial Join	1.14	1	1.3	**Spatial Manipulation**	11
Projection	1.14	1	1.3	**Spatial Transformation**	7
Area		1	1.3	**Spatial Calculation**	7
Math Formula	1.16	1	1.3	Manipulation	96
Multiple Ring Buffer	1.17	1	1.3	**Spatial Manipulation**	1
Overlay	1.17	1	1.3	**Spatial Manipulation**	3

Nodes	Debut Appearance			Category[17]	Frequency
	Figure	Case Study	Section		
Dissolve	1.19	1	1.3	**Spatial Manipulation**	7
GroupBy	1.20	1	1.3	Manipulation	58
Joiner	1.21	1	1.3	Manipulation	69
Geometry To Point	1.22	1	1.3	**Spatial Conversion**	8
Line Plot (JavaScript)	1.23	1	1.3	Views	3
Image Writer (Port)	1.25	1	1.3	IO	1
CSV Reader	2.2	2A1	2.1	IO	18
String Manipulation	2.3	2A1	2.1	Manipulation	8
Geocoding	2.4	2A1	2.1	**Spatial Conversion**	1
Lat/Lon to Geometry	2.5	2A1	2.1	**Spatial Conversion**	4
Concatenate	2.6	2A1	2.1	Manipulation	10
Coordinates XYZ		2A1	2.1	**Spatial Calculation**	5
Cross Joiner		2A1	2.1	Manipulation	2
GeoFile Writer		2A1	2.1	**Spatial IO**	3
Euclidean Distance	2.8	2A2	2.1	**Spatial Manipulation**	10
OSRM Distance Matrix	2.8	2A2	2.1	**Spatial Network**	1
Google Distance Matrix	2.8	2A2	2.1	**Spatial Network**	1
Road Network Distance Matrix	2.8	2A2	2.1	**Spatial Network**	6
Column Renamer	2.9	2A2	2.1	Manipulation	6
Scatter Plot (JavaScript)	2.10	2A2	2.1	Views	1
Linear Correlation	2.12	2A2	2.1	Analytics	2
Linear Regression Learner	2.13	2A2	2.1	Analytics	29
Round Double	2.14	2A2	2.1	Manipulation	3
CSV Writer	2.15	2A2	2.1	**Spatial IO**	8
Moving Aggregator	2.18	2B	2.2	Other Data Type	2
Rule-based Row Filter	3.3	3A	3.1	Manipulation	15
Double Configuration	3.5	3A	3.1	Workflow Abstraction	9
Spatial Heatmap	3.6	3A	3.1	**Spatial Visualization**	3
Kepler.gl Geoview(Time series points)[18]	3.11	3C	3.3	**Spatial Visualization**	7
Create Grid		3C	3.3	**Spatial Manipulation**	3

(Continued)

TABLE 1.4
Continued

Nodes	Debut Appearance			Category[17]	Frequency
	Figure	Case Study	Section		
Create Date&Time Range	3.12	3C	3.3	Other Data Type	1
Date&Time to String	3.13	3C	3.3	Other Data Type	1
Missing Value	3.14	3C	3.3	Manipulation	6
Table to R	3.17	3C	3.3	Scripting	3
R to Table	3.17	3C	3.3	Scripting	5
Voronoi (Thiessen) Polygon	4.2	4A	4.1	**Spatial Manipulation**	1
Nearest Join	4.2	4A	4.1	**Spatial Manipulation**	4
Kepler.gl Geoview (OD-flow 2D/3D)	4.6	4B1	4.2	**Spatial Visualization**	?
Spatial Weights	4.7	4B1	4.2	**Exploratory Spatial Data Analysis**	3
Python Script		4B1	4.2	Scripting	14
Component (Parameters)	4.15	4B2	4.2		
Column Selection Configuration	4.12	4B2	4.2	Workflow Abstraction	11
Single Selection Configuration	4.13	4B2	4.2	Workflow Abstraction	3
Boolean Configuration	4.14	4B2	4.2	Workflow Abstraction	1
Merge Variables		4B2	4.2	Workflow Control	3
Component Input		4B2	4.2		5
Component Output	4.15	4B2	4.2		5
Rule Engine	5.3	5B	5.2	Manipulation	12
Counting Loop Start	5.4	5C	5.3	Workflow Control	2
Math Formula (Variable)		5C	5.3	Workflow Control	2
String Manipulation (Variable)		5C	5.3	Workflow Control	1
Table Creator	5.5	5C	5.3	IO	1
Table Column to Variable		5C	5.3	Workflow Control	2
RowID	5.6	5C	5.3	Manipulation	12
Column Renamer	5.7	6C	6.3	Manipulation	6
Column Filter	5.8	7C	7.3	Manipulation	18
Component(2SFCA)	5.9	5C	5.3	2SFCA	

Nodes	Debut Appearance			Category[17]	Frequency
	Figure	Case Study	Section		
Column Rename (Regex)	5.10	5C	5.3	Manipulation	1
Loop End (Column Append)		5C	5.3	IO	1
Statistics		5C	5.3	Analytics	1
Kepler.gl Geoview (Multiple Layer)	6.2	6A	6.1	Spatial Visualization	7
Pivot	6.4	6B	6.2	Manipulation	7
Regression Predictor	6.5	6B	6.2	Analytics	1
Numeric Scorer	6.6	6B	6.2	Analytics	1
Table Transposer		6B	6.2	Manipulation	5
Column Appender		6B	6.2	Manipulation	8
Extract Table Dimension		6B	6.2	Manipulation	1
Table Row to Variable	6.7	6B	6.2	Workflow Control	3
Loop End	6.8	6B	6.2	Workflow Control	3
Normalizer	7.2	7	7.1	Manipulation	3
PCA Compute		7	7.1	Analytics	2
PCA Apply	7.3	7	7.1	Analytics	2
Distance Matrix Calculate	7.4	7	7.2	Analytics	1
Hierarchical Clustering (DistMatrix)	7.5	7	7.2	Analytics	1
Hierarchical Cluster View		7	7.2	Analytics	1
Hierarchical Cluster Assigner	7.6	7	7.2	Analytics	1
k-Means	7.8	7	7.2	Analytics	1
GeoPackage Reader		8A	8.1	Spatial IO	8
Table Row To Variable Loop Start		8A	8.1	Workflow Control	2
Rule-based Row Splitter		8A	8.1	Manipulation	1
Mean Center		8A	8.1	**Spatial Clustering**	2
Standard Deviational Ellipse		8A	8.1	**Spatial Clustering**	2
Polygon To Line		8A	8.1	**Spatial Conversion**	2
Add Table To R		8A	8.1	Scripting	3

(Continued)

TABLE 1.4

Continued

Nodes	Debut Appearance			Category[17]	Frequency
	Figure	Case Study	Section		
Number To String		8A	8.1	Manipulation	2
WKT to Geometry		8A	8.1	**Spatial Conversion**	1
Component		8A	8.1	Case Column	1
Global Getis-Ord G		8C	8.3	**Exploratory Spatial Data Analysis**	1
Global Moran's I		8C	8.3	**Exploratory Spatial Data Analysis**	1
Local Getis-Ord G		8C	8.3	**Exploratory Spatial Data Analysis**	1
Local Moran's I		8C	8.3	**Exploratory Spatial Data Analysis**	1
OLS with Spatial Test		8C	8.3	**Spatial Modelling**	1
GWR Model	8.9	8C	8.3	**Spatial Modelling**	1
Component	8.10	8C	8.3	Visualization Layout	1
SKATER	9.2	9A	9.1	**Spatial Clustering**	1
SCHC		9A	9.1	**Spatial Clustering**	1
REDCAP		9A	9.1	**Spatial Clustering**	1
AZP-Greedy		9A	9.1	**Spatial Clustering**	1
Peano Curve		9A	9.1	**Spatial Clustering**	3
MSSC Initialization	9.2	9A	9.1	**Spatial Clustering**	3
MSSC Refiner	9.2	9A	9.1	**Spatial Clustering**	3
MaxP-Greedy		9A	9.1	**Spatial Clustering**	1
Column Aggregator		9B	9.2	Manipulation	3
Math Formula (Multi Column)	9.6	9B	9.2	Manipulation	8
Length		9B	9.2	**Spatial Calculation**	1
Convex Hull		9B	9.2	**Spatial Calculation**	1
Bounding Circle		9B	9.2	**Spatial Calculation**	1
Bounding Box		9B	9.2	**Spatial Calculation**	1
Bounds		9B	9.2	**Spatial Calculation**	1
Row Splitter		9C	9.3	Manipulation	3
Component (MSSC)	9.8	9C	9.3		
Component (DissJoin)	9.8	9C	9.3		
Isolation Tackler		9C	9.3	**Spatial Clustering**	2

Nodes	Debut Appearance			Category[17]	Frequency
	Figure	Case Study	Section		
Line Plot (Plotly)	10.2	10	10.3	JavaScript View(labs)	1
R To R		11A	11.1	Scripting	1
P-median Solver	11.5	11B2	11.2	**Location Analysis**	1
MCLP Solver	11.5	11B2	11.2	**Location Analysis**	1
Duplicate Row Filter	11.6	11B2	11.2	Manipulation	1
Reference Column Filter		11B2	11.2	Manipulation	1
P-center Solver	11.5	11B2	11.2	**Location Analysis**	1
Component (Gravity2SFCA)		11B3	11.2		
MAEP Solver	11.8	11B3	11.2	**Location Analysis**	1
Parallel Chunk Start	12.3	12A	12.1	Parallel Execution	2
Create Random Points	12.2	12A	12.1	**Spatial Transformation**	3
Parallel Chunk End	12.3	12A	12.1	Parallel Execution	2
Multipart To Singlepart		12A	12.1	**Spatial Transformation**	3
Bar Chart (JavaScript)	12.4	12A	12.1	Views	1
Sorter	12.8	12B	12.2	Manipulation	1
Local File Browser Configuration	13.2	13	13.1	Workflow Abstraction	1
Python View		13	13.1	Scripting	1
Component (ABM simulation)		13	13.1		
String to Date&Time	14.2	14A	14.1	Other Data Type	1
Rank	14.3	14A	14.1	Manipulation	1
Points To Line	14.4	14A	14.1	**Spatial Transformation**	4
Reference Row Filter		14A	14.1	Manipulation	1
GeoPackage Writer	14.6	14A	14.1	Spatial IO	3

used nodes are Math Formula (96), Joiner (69), GroupBy (58), GeoFile Reader (53), Linear Regression Learner (29), Geospatial View (25), Column Filter (18), CSV Reader (18), and Row Filter (16).

Figure 1.27 displays the frequency of the top 45 nodes. Besides nodes for reading and visualizing geospatial data, Spatial join, Euclidean distance, Projection and Dissolve are the most commonly used in the Geospatial Analytics Extension for KNIME in this Lab Manuel.

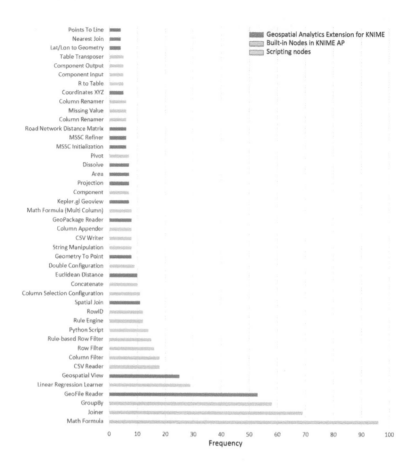

FIGURE 1.27   Frequency of Top 45 Nodes Used in KNIME Lab Manual

NOTES

1   In the classical UI, all nodes are shown in a list, and users can select the nodes category to explore or compare node communities. In the Modern UI, all nodes are shown as big icons, and users may rely on Search Nodes to find one and thus it is convenient for experienced users.
2   To display RAM usage, go to File > Preference > General, check Show heap status.
3   This setting is consistent with that for the case studies based on ArcGIS Pro, suitable for the installation of *arcgisbinding package*.
4   https://docs.knime.com/latest/python_installation_guide/#bundled-env
5   Go to GitHub for latest packages: https://github.com/spatial-data-lab/knime-geospatial-extension.
6   User can also access these data on GitHub: https://github.com/UrbanGISer/CGA-KNIME-Workbook
7   https://raw.githubusercontent.com/UrbanGISer/CGA-KNIME-Workbook/main/data/chp1/County.zip
8   The default renderers for Geospatial column can be changed via Preference > KNIME >

KNIME > Preferred Renders.

9   https://raw.githubusercontent.com/UrbanGISer/CGA-KNIME-Workbook/main/data/chp1/Tract.zip

10 Users may refer to Shapely documentation, https://shapely.readthedocs.io/en/latest/manual.html#binary-predicates

11 Visit the EPSG website for more information: https://epsg.io/26915

12 https://raw.githubusercontent.com/UrbanGISer/CGA-KNIME-Workbook/main/data/chp1/BRcenter.zip

13 If uncheck the option "Return only geometries of the same geometry type", the resulting output may include different types of geometries. For example, when two lines intersect, points may be generated as part of the output.

14 https://raw.githubusercontent.com/UrbanGISer/CGA-KNIME-Workbook/main/data/chp1/BRblk.zip

15 As all PopuDen with the same Distance are identical, other aggregation choices such as first, max, mean, and median work as well.

16 As the plot is vector-based in a SVG format, the image size only affects the font sizes and labels of the output image. The SVG format is a popular choice for displaying 2D graphics, charts, and illustrations on websites, and can be scaled up or down without losing any resolution.

17 Nodes in Geospatial Analytics extension are highlighted with Bold font.

18 The frequency count records the number of times a node appears with different usages, such as the Component and Kepler.gl Geoview.

2 Measuring Distance and Time and Analyzing Distance Decay Behavior

This chapter uses two case studies to illustrate how to implement two common tasks encountered most often in spatial analysis: estimating a travel distance or time matrix and modeling distance decay behaviors. Case Study 2A shows how to measure distances and travel times between residents at the census block group level and acute hospitals in Baton Rouge, Louisiana. Case Study 2B uses hospitalization data in Florida to demonstrate how to derive the best fitting distance decay function by the spatial interaction model or the complementary cumulative distribution curve.

2.1 CASE STUDY 2A: ESTIMATING TRAVEL TIMES TO HOSPITALS IN BATON ROUGE

This case study illustrates how to estimate various travel time matrices between census block groups and hospitals in East Baton Rouge Parish (EBRP), hereafter simply referred to as Baton Rouge, Louisiana. The census block group data in the study area is based on the 2020 Census as explained in Chapter 1. The hospitals data is extracted from the membership directory of Louisiana Hospital Association (2021). Results from this project will be used in Case Study 4A of Chapter 4 that delineates hospital service areas by the Huff model, and in Case Study 5 of Chapter 5 to measure spatial accessibility of hospitals.

The following data sets are provided under the data folder `BatonRouge`:

1. a comma-separated text file `Hosp_Address.csv` containing addresses, geographic coordinates, and numbers of staffed beds for the five acute hospitals in EBRP,
2. a zipped ESRI Shapefile feature class `BR_Bkg.zip` for the census block group data, a feature class `BR_MainRd.zip` for main roads, all in EBRP,
3. two ESRI Shapefiles `hosp.shp` and `BR_BkgPt.shp` generated by Step 8 in Section 2.1 for geocoding data from `Hosp_Address.csv`, and centroids of census block group data, respectively, and
4. a CSV file `OD_Drive_Time.csv` generated by Step 12 in Section 2.1 for distance matrix between the census block centroids and hospitals.

Figure 2.1 shows the workflow for this case study.

DOI: 10.1201/9781003304357-2

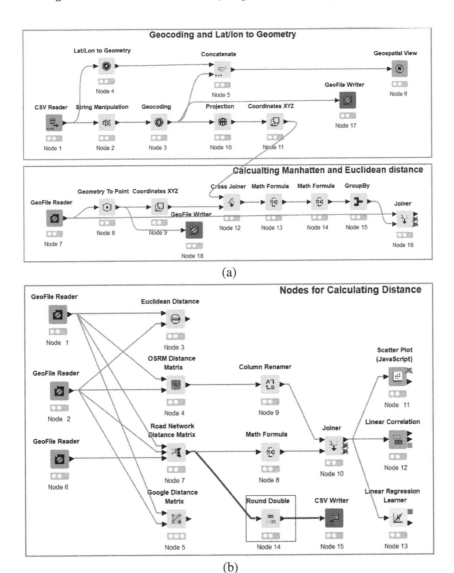

(a)

(b)

FIGURE 2.1   Workflow of Case Study 2A

2.1.1   GEOCODING HOSPITALS FROM STREET ADDRESSES
OR GEOGRAPHIC COORDINATES

Step 1. Preparing the address data: Create a new workflow Case2A1. Under Node Repository > 📥 IO > 🗋 Read, drag 🔲 CSV Reader to the Editor canvas. In the dialog of CSV Reader (Node 1), (1) under the Settings tab[1] (Figure 2.2a), click Browse to choose the file Hosp_Address.csv, then click Autodetect format to format

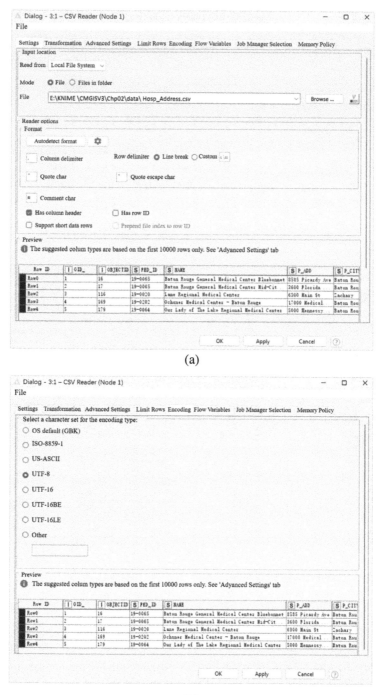

(a)

(b)

FIGURE 2.2 Interface of CSV Reader: (a) Setting tab, and (b) Encoding tab

FIGURE 2.3   Interface of String Manipulation

the table automatically; (2) under the Encoding tab (Figuire 2.2b), check `UTF-8`, and verify the table under Preview box to ensure that it is in the desired format. Click OK to execute the node. The resulting table contains three address columns (`P_ADD`, `P_CITY`, and `P_STATE`) and two coordinate columns (`LATITUDE`, `LONGITUDE`).

Step 2. Processing addresses by String manipulation: String Manipulation is used to concatenate the three address columns and generate a full address for geocoding.

Under 🔧 Manipulation > ▌ Column > 🗔 Convert & Replace), drag ʀ[s] String Manipulation to the Editor canvas, link CSV Reader (Node 1) to it. In the dialog of String Manipulation (Node 2) (Figure 2.3), (1) under String Manipulation tab, input `joinSep(",",$P_ADD$,$P_CITY$,$P_STATE$,"USA")` in the Expression box by double-click on `joinSep(sep,str . . .)` under Function and the column names `P_ADD`, `P_CITY`, `P_STATE` under Column List; (2) Check Append Column, and input `FullAddress`. Click OK and execute the node.

The resulting table contains a column `FullAddress` for geocoding. Users can refer to the description and examples for all functions in String Manipulation.

Step 3. Geocoding by addresses and applying Lat/Lon to Geometry: Under ⊕ Geospatial Analytics > ⬤ Spatial Conversion, drag ⊙ Geocoding to the Editor canvas, and link String Manipulation (Node 2) to it. In the dialog of Geocoding (Node 3) shown in Figure 2.4, choose `FullAddress` for Address column, `nominatim` for the Service provider, leave the API Key value empty, and keep other default settings. Click OK and execute the node.

The node appends a point geometry column to the input table. The default geocoding service Nominatim is free but with speed limitation. Some other service providers may require an API key. Refer to the node's description for more detail.

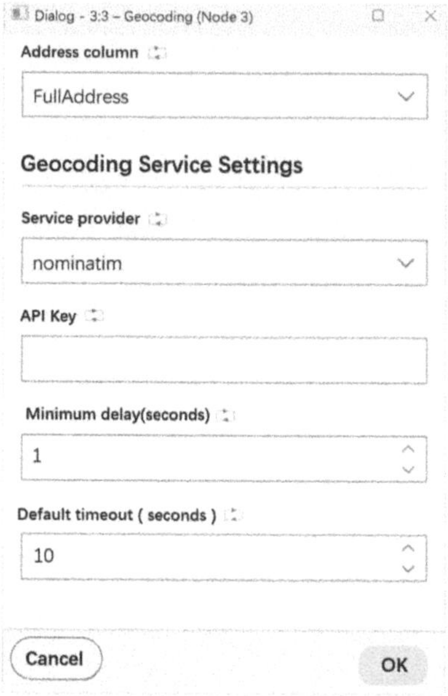

FIGURE 2.4   Interface of Geocoding

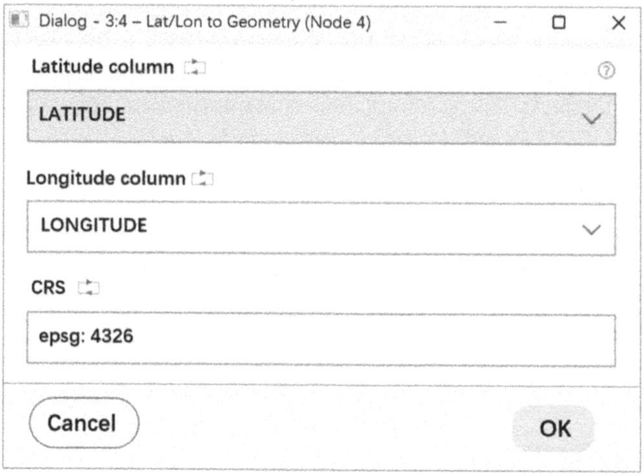

FIGURE 2.5   Interface of Lat/Lon to Geometry

Another alternative is to convert geographic coordinates to geometry data.

Under ⬤ Spatial Conversion, drag ⬤ Lat/Lon to Geometry to the Editor canvas, and link CSV Reader (Node 1) to it. In the dialog of Lat/Lon to Geometry (Node 4) shown in Figure 2.5, choose LATITUDE for Latitude column, LONGITUDE for Longitude column, and keep the default value epsg:4326 for CRS. Click OK and execute the node. This node appends a geometry column based on latitude and longitude.

Step 4. Optional: Comparing geocoded results: In Node Repository > 🔧 Manipulation > ▬ Row > 🗋 Transform, drag 🔧 Concatenate to the Editor canvas,

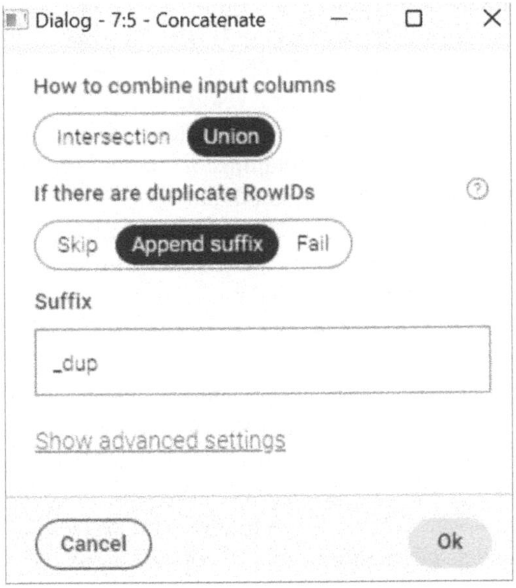

FIGURE 2.6   Interface of Concatenate

and link Lat/Lon to Geometry (Node 4) and Geocoding (Node 3) to its two input ports. Execute Concatenate (Node 5) directly using the default settings shown in Figure 2.6.

Add a Geospatial View and link Concatenate (Node 5) to it. In the Geospatial View dialog (Node 6), set the Marker size scale to 10 and keep the other default settings. Execute it. The visualization shows that the two sets of geocoded results are similar with minor differences. The result from address geocoding will be used for the remainder of the case study.

2.1.2   ESTIMATING EUCLIDEAN AND MANHATTAN DISTANCES BASED ON THE OD COORDINATES

Step 5. Generating projected coordinates of hospitals and census block groups: Under Geospatial Analytics > Spatial IO, drag GeoFile Reader to the Editor canvas. In the dialog of GeoFile Reader (Node 7), input the file path of BR_Bkg.zip for Census block group data in Baton Rouge, e.g., E:\KNIME\CMGISV3\Chp02\ data\BR_Bkg.zip, or input a URL path.[2] Execute it.

Under Geospatial Transformation, add a new node Geometry to Point (Node 8) and link GeoFile Reader (Node 7) to it. Set centroid for Point Type Selection, check Replace for Output column in its dialog and execute it. The centroids of census block groups are used to calculate the distances.

Under 🌐 Spatial Calculation, drag 🗺 Coordiantes XYZ to the Editor, and link Geometry to Point (Node 8) to it. Execute Coordiantes XYZ (Node 9) with the default settings. It attaches two columns, x and y, representing its longitude and latitude (already in projected units).

The hospital data with a geographic coordinates reference system (epsg:4326) needs to be transformed into projected coordinates. Add two new nodes, Projection (Node 10) and Coordiantes XYZ (Node 11), and link Geocoding (Node 3)→ Projection (Node 10) → Coordiantes XYZ (Node 11). In the Projection dialog, input `epsg:26915` for New CRS, and keep the default settings for Coordinates XYZ. Execute the two nodes. Similarly, two columns for longitude and latitude (x and y) in projected units are attached to the hospital data.

Step 6. Calculating Euclidean and Manhattan distances by formulas: To generate a list of origin-destination distances, we will need to perform a cross join on the hospital and census block group centroid data.

In Node Repository > 🔧 Manipulation > ▮ Column > 🗂 Split& Combine, drag 🔀 Cross Joiner to the Editor canvas, and link the two Coordiantes XYZ nodes (Nodes 11 and 9) to the two input ports of Cross Joiner (Node 12). Execute the node directly with the default settings. The resulting table has 325×5=1620 records, and any duplicated column names are automatically updated with a suffix of "(#1)", such as x(#1) and y(#1).

Under ▮ Column > 🗂 Convert& Replace, drag two Math Formula nodes to the Editor, and link Cross Joiner (Node 12) → Math Formula (Node 13) → Math Formula (Node 14). In the dialog of Math Formula (Node 13), input `(($x$- $x (#1)$)^ 2+($y$-$y (#1)$)^ 2)^0.5` in the Expression by using the Column List and Function options to select the variable names and operation symbols, check `Append Column`, and name the new column `EuclideanDist`. For Math Formula (Node 14), input `abs($x$- $x (#1)$)+abs($y$-$y (#1)$)` for the Expression and check `Append Column`, and name the new column `ManhDist`.

These two nodes calculate the Euclidean and Manhattan distances based on the projected coordinates, respectively.

Step 7. Mapping the shortest Euclidean distances to hospitals: In Node Repository > 🔧 Manipulation > ▬ Row > 🗂 Transform, drag ⊃ GroupBy to the Editor, and link Math Formula (Node 14) to it. In the dialog of GroupBy (Node 15) (refer to Step 11 in Section 1.3), (1) under the default tab Settings and its sub-tab Groups, choose `OBJECTID(#1)` for Group column(s); (2) under Advanced settings, choose `Keep original name(s)` for Column naming; (3) under Manual Aggregation tab, choose `EuclideanDist` and `ManhDist` for Column, minimum for Aggregation, and keep the other default settings. Click OK and execute the node. This node calculates the shortest distance for each census block group based on its unique `OBJECTID(#1)`.

In Manipulation > ▮ Column > 🗂 Split& Combine, drag ⅄ Joiner to the Editor canvas, link GeoFile Reader (Node 7) to the top input port of Joiner (Node 16), and link GroupBy (Node 15) to the bottom port of Joiner (Node 7). In the dialog (refer to Step 7 in Section 1.2), under the Joiner Settings tab, (1) for Join Columns, keep all of the following checked for Match, (2) click Add matching criterion, and choose `OBJECTID` for Top Input ('Left' table) and `OBJECTID(#1)` for Bottom Input ('right' table), (3) under Column selection tab, keep all fields in the Include box for Top Input (left table), keep `EuclideanDist` and `ManhDist` in the Include box for Bottom Input(right table), and (4) keep other default setting. Click OK and execute it.

Link Joiner (Node 16) to Geospatial View (Node 6) to visualize the shortest distance for each hospital by choosing `EuclideanDist` or `ManhDist` for Color map (Figure 2.7).

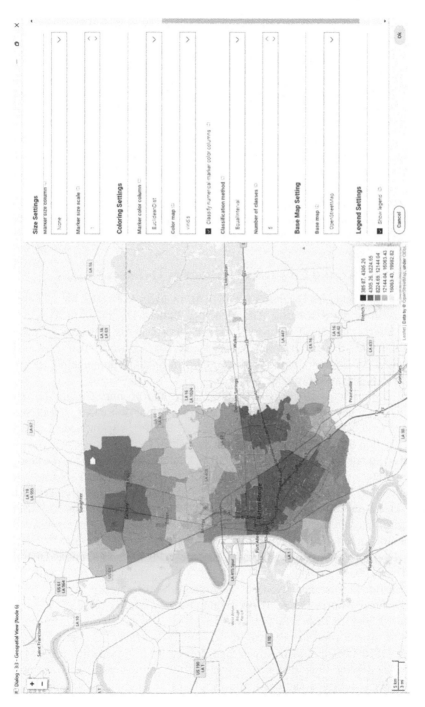

FIGURE 2.7   Euclidean distances to the nearest hospitals in Baton Rouge

Step 8. Exporting the result of geospatial data by GeoFile Writer nodes: To export the geocoding data for hospitals and the census block group centroids, there are a couple of options in KNIME. One way is to use the GeoFile Reader node to save the data as a single file, while another way is to use the GeoPackage Reader node to save it as a layer in a GeoPackage file. Here we use the GeoFile Reader node as an example.

Under ⬛ Spatial IO, drag ⬛ GeoFile Writer to the Editor, link Geocoding(Node 3) to the node GeoFile Writer(Node 17). In the dialog of GeoFile Writer(Node 17), check Shapefile and name the output file as `hosp.shp` by inputting a full path, e.g., `E:\KNIME\CMGISV3\Chp02\data\hosp.shp`. Click OK and execute the node to save the data. The other 4 associated files with the same name as `hosp.shp` but with different extensions will be saved as well, such as `hosp.prj`, `hosp cpg`, `hosp.shx`, and `hosp.dbf`.

Add another GeoFile Writer node (Node 18) and link Geometry To Point (Node 8) to it. Configure it to save the data as `BR_BkgPt.shp` with a full path, e.g., `E:\KNIME\CMGISV3\Chp02\data\BR_BkgPt.shp`.

2.1.3 ESTIMATING AND COMPARING DISTANCES BY DIFFERENT SOURCES

The Geospatial Analytics Extension for KNIME includes five nodes for calculating travel costs in origin-destination (OD) matrices. OSRM Distance Matrix, Google Distance Matrix and Road Network Distance Matrix in the category of Spatial Network calculate the distance matrix based on a road network. Two other nodes for distance calculation are based on coordinates in Spatial Manipulation category: Euclidean Distance and Haversine Distance.[3]

Step 9. Calculating with Euclidean Distance, OSRM and Google Distance Matrix: Create a new workflow `Case2A2`. Add two GeoFile Reader nodes (Nodes 1 and 2) to read census block group centroids `BR_BkgPt.shp` in Node 1, and the hospital location data `hosp.shp` in Node 2, respectively.

Add Euclidean Distance (Node 3), OSRM Distance Matrix (Node 4), Google Distance Matrix (Node 5) to the Editor, and link GeoFile Reader (Node 1) and GeoFile Reader (Node 2) to the two input ports of those nodes, respectively. In the dialog, set the parameters as shown in Figure 2.8a–c, e.g., choosing `OBJECTID` for Origin ID column, `OID_` for Destination ID column, `Kilometer` for Distance unit in, checking `Travel Cost` for Result mode in OSRM Distance Matrix (Node 4), or inputting a valid Google API Key[4] in Google Distance Matrix (Node 5).

The Euclidean Distance (Node 3) generates a similar result as Step 6. The OSRM Distance Matrix (Node 4) or Google Distance Matrix (Node 5) generates an OD travel cost list table with four columns: `Origin ID`, `Destination ID`, `Duration` (travel time in minutes) and `Distance` (travel distance in meters). The node OSRM Matrix utilizes the Open-Source Routing Machine (OSRM)[5] to calculate the driving time, distance, or route of the shortest paths between origin (left or upper input port) and destination (right or lower input port) points based on the open and free road network data of the OpenStreetMap (OSM) project. The

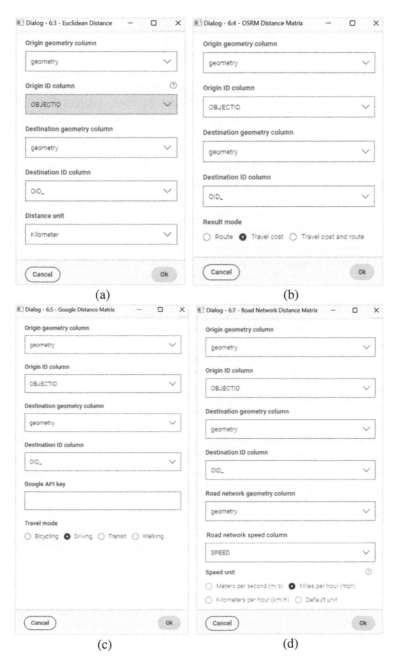

FIGURE 2.8   Interfaces of nodes for OD distance matrix calculation: (a) Euclidean Distance, (b) OSRM Distance Matrix, (c) Google Distance Matrix and (d) Road Network Distance Matrix

node Google Distance Matrix yields a travel distance and time matrix based on the Google Maps recommended routes between start and end points and offers four travel modes: bicycling, driving, transit and walking.

If the input geometry is not a point feature, centroids are used. The output includes numerical indices for the origin and destination data that can serve as a common key for merging the data.

Step 10. Calculating Road Network Distance Matrix: The Road Network Distance Matrix node calculates the travel time and distance between origin (left or upper input port) and destination (right or lower input port) points based on a user-specified road network. It first snaps the origin and destination points to the road network and uses the function *single_source_dijkstra_path_length* in Python package *NetworkX* to compute the shortest path length between the source and all other reachable nodes for the weighted (time or distance) graph. The calculation depends on the projected coordinate system (CRS) of the road network datasets. If the dataset does not have a projected CRS, a default project CRS based on UTM zone will be applied. Similarly, if the input geometry is not a point feature, the centroids are used. Chapter 12 will show the Python code of calculating the shortest paths based on a road network.

Add a GeoFile Reader node (Node 6). In the dialog, input the file path for road network e.g., `E:\KNIME\CMGISV3\Chp02\data\BR_MainRd.zip`. Click OK and execute it to read the road data.

Add a Road Network Distance Matrix node (Node 7). Link three GeoFile Reader nodes (Nodes 1, 2 and 6) to the three input ports of Node 7. In the dialog of Node 7 as shown in Figure 2.8d, choose `OBJECTID` for Origin ID column, `OID_` for Destination ID column, SPEED for Road network speed column and check `Miles per hour(mph)` for Speed unit. Click OK and execute it.

Compared to OSRM Distance Matrix (Node 4) and Google Distance Matrix (Node 5), the result table of Road Network Distance Matrix node contains two additional columns, `Origin snap distance` and `Destination snap distance`, which are the distances between the origin and destination points and the corresponding nearest points on the road network, respectively, in meters. Users may choose to ignore or include these values in the total distances according to their research purpose.

Add a Math Formula node (Node 8) and link Road Network Distance Matrix (Node 7) to it. Configure the node to append a new column `Distance1` for the total distance with the expression

`$Distance$+$Origin    snap    distance$+$Destination    snap    distance$`.

Before joining and comparing the travel times by OSRM Distance Matrix and Road Network Distance Matrix, we need to rename the columns of one of the tables as they share the same names.

Step 11. Joining Road Network Distance Matrix and OSRM Matrix results: Under
🪄 Manipulation > ▮ Column > ▢ Convert & Replace), drag ⟁ Column Renamer[6] to the Editor canvas, and link OSRM Distance Matrix (Node 4) to it. In the dialog of Column Renamer (Node 9) shown in Figure 2.9, click Add column to add two lines of renaming setting boxes, then choose the variable names `Distance` and `Duration`

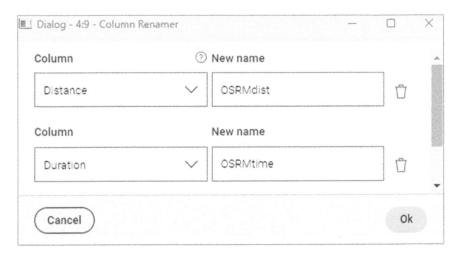

FIGURE 2.9   Interface of Column Renamer

in the left boxes and rename them to OSRMdist and OSRMtime, respectively in the right boxes. Click OK and execute the node.

We can also use flow variables to specify column names in the Column Renamer node. By right-clicking the node and selecting Configure Flow Variables . . . , users can choose flow variables as column names, which can facilitate workflow automation. This technique will be practiced extensively in Chapter 5.

Add a Joiner node (Node 10), and link Column Renamer (Node 9) and Math Formula (Node 8) to the two input ports of Joiner (Node 10). In the dialog of Joiner (Node 10) (refer to Step 7 in Section 1.2), follow these steps: (1) under Joiner Setting tab, add two conditions under Join columns by clicking Add matching criterion and + button to add two joining conditions, set Origin ID (or Destination ID) for both Top Input ('left' table) and Bottom Input ('right' table); (2) under Column Selection, exclude Origin ID and Destination ID in the Include box for either Top input (left table) or Bottom input (right table) to avoid duplicate columns; and (3) keep other default settings. Click OK and execute the node.

Step 12. Applying Scatter Plot, Linear correlation, and regression: Under Q Views > ᴊꜱ JavaScript, drag ⠿ Scatter Plot to the Editor Canvas, and link Joiner (Node 10) to it. This node shares most of the parameters as Line Plot (Refer to Step 12 in Section 1.3). In the dialog of Scatter Plot (Node 11) shown in Figure 2.10, (1) under Options tab, check Create image at output, select Duration for Choose column for x axis and OSRMtime for Choose column for y axis; (2) under Axis Configuration, input Duration by Road Network (Minutes) and Duration by OSRM (Minutes) for Labels; (3) under General Plot Options tab, revise the Width of image and Height of image as 600 and 400, respectively; and (4) keep other default settings. Click OK and execute it.

FIGURE 2.10   Interface of Scatter Plot

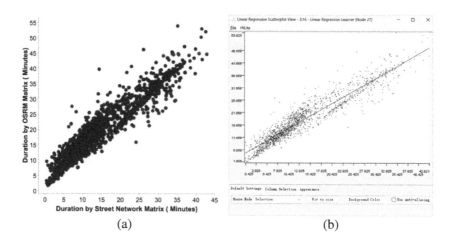

FIGURE 2.11   Scatter Plot on distance comparison: (a) Scatter Plot Node, and (b) View: Linear Regression Scatterplot View

Right click Scatter Plot (Node 11), choose ☌ Interactive View: Line Plot to view the plot, or choose Image to view the output image (Figure 2.11a).

Under 🌐 Analytics > Σ Linear Correlation, drag ▦ Linear Correlation to the Editor Canvas, and link Joiner (Node 10) to it. In the dialog of Linear Correlation (Node 12) shown in Figure 2.12, choose Duration, Distance, OSRMtime, and OSRMdist in the Include box, and keep other default settings. Click OK and execute

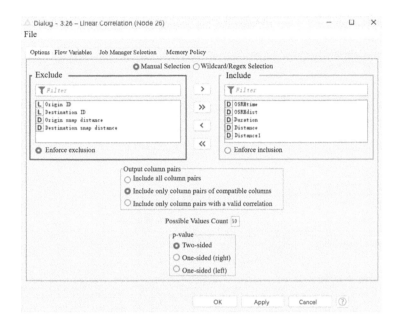

FIGURE 2.12 Interface of Linear Correlation

the node. This node has four output results on its right-click menu View: Correlation Matrix (as shown in Figure 2.16a), Correlation measure and Correlation model.

Under 🔵 Analytics > 🔵 Mining > ⬜ Linear/Regression Learner, drag ⬛ Linear Regression Learner to the Editor Canvas, and link Joiner (Node 10) to it. In the dialog of Linear Regression Learner (Node 13) shown in Figure 2.13, under Options tab, choose OSRMtime for Target, set Duration in the Include box, and keep other default settings. Click OK and execute it. This node has four outputs: (1) View: Linear Regression Result View, (2) View: Linear Regression Scatterplot View (Figure 2.11b), (3) Model for Predictor, and (4) Coefficients and Statistics. The last two items can be accessed via its output ports.

Step 13. Exporting distance matrix result table: Here we will export the result of Road Network Distance Matrix with rounding data, as it may be desirable to limit the digits in the result table. Under 🔧 Manipulation > ▮ Column > ⬜ Convert & Replace, drag ≈ Round Double to the Editor canvas, and link Road Network Distance Matrix (Node 7) to it. In the dialog of Round Double (Node 14) shown in Figure 2.14, uncheck Append as new columns (specify suffix), and keep other default settings (Double for Output format, 3 for Precision, Decimal places for Precision mode, HALF_UP for Rounding mode). Click OK and execute the node.

Under Node Repository > 📥 IO > ⬡ Write, drag 🖥 CSV Writer to the Editor canvas, and link Round Double (Node 14) to it. In the dialog of CSV Writer (Node 15) shown in Figure 2.15, (1) under the Settings tab, click Browse to choose the

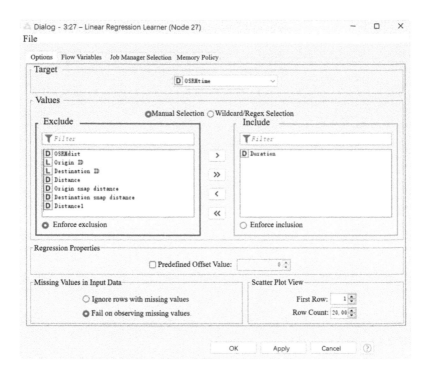

FIGURE 2.13   Interface of Linear Regression Learner

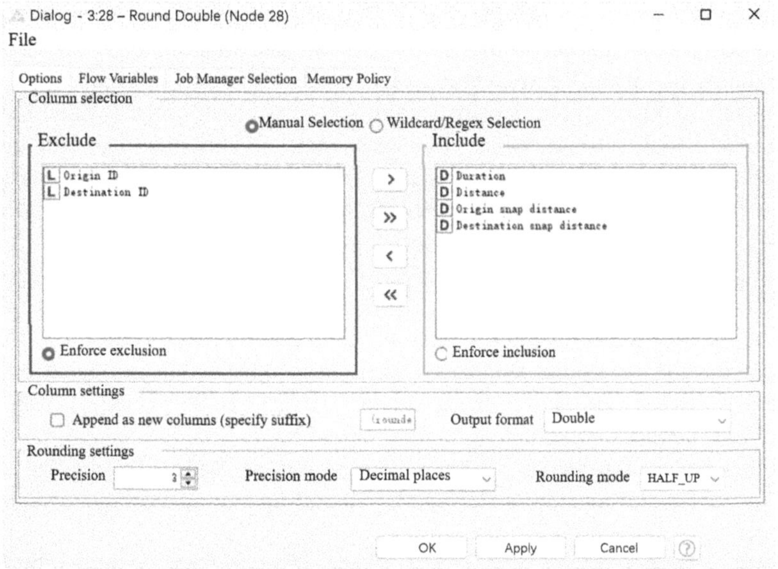

FIGURE 2.14   Interface of Round Double

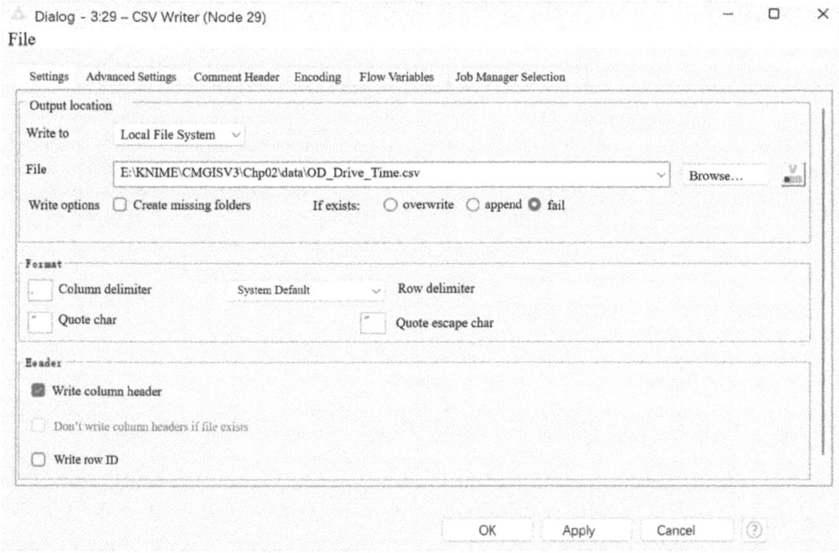

FIGURE 2.15 Interface of CSV Writer

save path, e.g., `E:\KNIME\CMGISV3\Chp02\data\OD_Drive_Time.csv`, and keep other default settings. Click OK and exectue it. One may revise the output file as a zipped file by checking `Compress output file (gzip)` in Advanced Settings, changing the save path format as `.csv.gz` and checking Compress output file(gzip) in the Advance Setting tab to compress the data, or revising the encoding type as `UTF-8` in the Encoding tab, etc.

The result of Linear Regression indicates that Road Network Distance Matrix underestimates the travel times by approximately 5 minutes compared to OSRM matrix, consistent with the result of related literature (Wang & Xu, 2011). As shown in Figure 2.16b, the correlation coefficient between the two duration estimates (0.945) is slightly lower than that of the two distance estimates (0.982).

2.2 CASE STUDY 2B: ANALYZING DISTANCE DECAY BEHAVIOR FOR HOSPITALIZATION IN FLORIDA

This case study is based on a project reported in Wang and Wang (2022), and examines the distance decay rule via two approaches, namely the spatial interaction model and the complementary cumulative distribution curve. The main data source is the State Inpatient Database (SID) in Florida in 2011 from the Healthcare Cost and Utilization Project (HCUP) sponsored by the Agency for Healthcare Research and Quality (AHRQ, 2011). The SID dataset includes individual inpatient discharge records from hospitals, which are aggregated to OD inpatient volumes between residential ZIP code areas and hospitals.

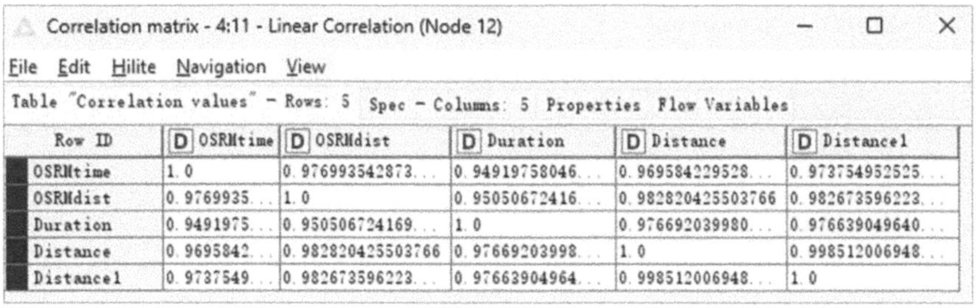

Row ID	D OSRMtime	D OSRMdist	D Duration	D Distance	D Distance1
OSRMtime	1.0	0.976993542873...	0.94919758046...	0.969584229528...	0.973754952525...
OSRMdist	0.9769935...	1.0	0.95050672416...	0.982820425503766	0.982673596223...
Duration	0.9491975...	0.950506724169...	1.0	0.976692039980...	0.976639049640...
Distance	0.9695842...	0.982820425503766	0.97669203998...	1.0	0.998512006948...
Distance1	0.9737549...	0.982673596223...	0.97663904964...	0.998512006948...	1.0

(a)

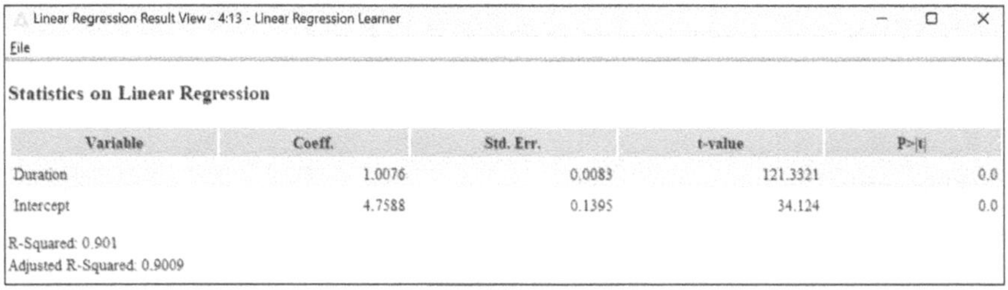

Variable	Coeff.	Std. Err.	t-value	P>|t|
Duration	1.0076	0.0083	121.3321	0.0
Intercept	4.7588	0.1395	34.124	0.0

R-Squared: 0.901
Adjusted R-Squared: 0.9009

(b)

FIGURE 2.16 (a) Correlations between various distance and duration measures, and (b) Regression on duration estimates

The case study uses an OD flow table `OD_All_Flows.csv` with 209,379 flows in the folder `Florida`, with the following fields (columns):

(1) `Hosp_ZoneID`, `Hosp_X_MC`, and `Hosp_Y_MC` for each hospital's ID and XY-coordinates, and `NUMBEDS` for its staffed bed size,
(2) `PatientZipZoneID`, `Xcoord`, and `Ycoord` for each ZIP code area's ID and XY-coordinates, and `POPU` for its population, and
(3) `AllFlows` and `Total_Time_Min` for patient service flow volume and drive time (minutes) on each OD pair.

Figure 2.17 summarizes the workflow for this case study.

Step 1. Preparing the OD flow list: Create a new workflow `Case2B`. Under Node Repository > IO > Read, drag CSV Reader to the Editor canvas. In the dialog of CSV Reader (Node 1) (refer to Figure 2.2a-b if needed), (1) under the Settings tab, click Browse to choose the zipped csv file `OD_All_Flows.csv.gz`, and click the button Autodetect format to format the table automatically; (2) under the Encoding tab, check `UTF-8`; (3) check the table under Preview box to make sure if it is in the desirable form. Click OK and execute the node.

Search in the box of Node Repository for Row Filter and add it to the Editor, and link CSV Reader (Node 1) to it. In the dialog of Row Filter (Node 2) (refer to step 4 in Section 1.2), choose `AllFlows` for Column to test, check `use range checking`, input `1` for lower bound, and check `Include rows by attribute value`. Click OK and execute it. This node includes all the records (n=37,180) with `AllFlows>0`.

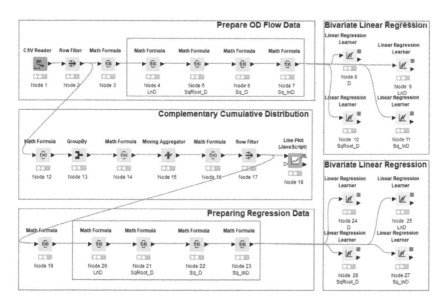

FIGURE 2.17 Workflow of Case Study 2B

Search and add 5 new Math Formula nodes, and link Row Filter (Node 2) to them sequentially. In the dialog of Math Formula (Node 3), check Append Colum, input LnI in the box, and calcualte it with the Expression ln($AllFlows$/ ($POPU$ * $NUMBEDS$)). Hereafter, this type of calculation will be rephrased as, "append a new column LnI (or update the column, if Replace column checked) with the expression ln($AllFlows$/ ($POPU$ * $NUMBEDS$))". Click OK and execute this node.

For the other four Math Formula nodes (Nodes 4–7), append four columns, LnD, SqRoot_D, Sq_D and Sq_lnD, with the following expressions in the order of,

```
ln($Total_Time_min$)
sqrt($TOTAL_TIME_MIN$)
$TOTAL_TIME_MIN$^ 2
$LnD$^ 2
```

Step 2. Implementing OLS Regression of the spatial interaction model: Add four new Linear Regression Learner nodes 8–11, and link Math Formula nodes 7 to the four nodes. In the dialog of Linear Regression Learner (for example, Node 8), chosoe LnI for Target, set Total_Time_min in the Include box for Values, and keep other default settings. Click OK and execute it.

For the other three nodes, set LnI as Target and use the following Values: LnD, SqRoot_D, and Sq_lnD, respectively.

Based on the regression results, the best-fitting distance decay function with the highest $R^2 = 0.443$ is the power function, lnI = -7.8397-1.4342lnd. This completes the estimation of distance decay functions by the spatial interaction model.

Step 3. Calculating flow volumes by 1-minute interval: Add a new Math Formula node (Node 12) and link Row Filter (Node 2) to it. In the dialog of Math Formula (Node 12), input `ceil($Total_Time_min$)` in expression, check `Replace column` and choose `Total_Time_min` for it, and check `Convert to Int`, which rounds a number up to the next largest integer.

Add a GroupBy node (Node 13) and link Math Formula (Node 12) to it. In the dialog of GroupBy (Node 13), under the Settings tab and its sub-tab Groups, (1) choose `Total_Time_min` for Group column(s); (2) under Advance setting, choose `Keep original name(s)` for Column naming; (3) under Manual Aggregation tab, choose `Allflows` under Column and `Sum` for its aggregation method. Click OK and execute the node. The column `AllFlows` in the result table represents the total flows aggregated by one-minute increment.

Step 4. Calculating complementary cumulative percentage of flows: Add a Math Formula node (Node 14), and link GroupBy (Node 13) to it. Configure it to append a new column `TotalFlows` as percentage of flows with the expression `$AllFlows$/ COL_SUM($AllFlows$)* 100`.

Under Node Repository > Other Data Types > ∿ Time series> ▢ Smoothing, drag ✦ Moving Aggregation to the Editor, and link Math Formula (Node 14) to it. In the dialog of Moving Aggregation (Node 15) shown in Figure 2.18, (1) under the Setting tab, check `Cumulative computation`, and uncheck the other two options `Remove aggregation column` and `Remove retained columns`; (2) under Aggregation settings, double click the column name `TotalFlows` under Available columns to select it in the right side box; (3) under the Column in the right side box, choose `Sum` for Aggregation that corresponds to column `TotalFlows`; and (3) keep other default settings (e.g., `Aggregation method(column)` for Column Naming).

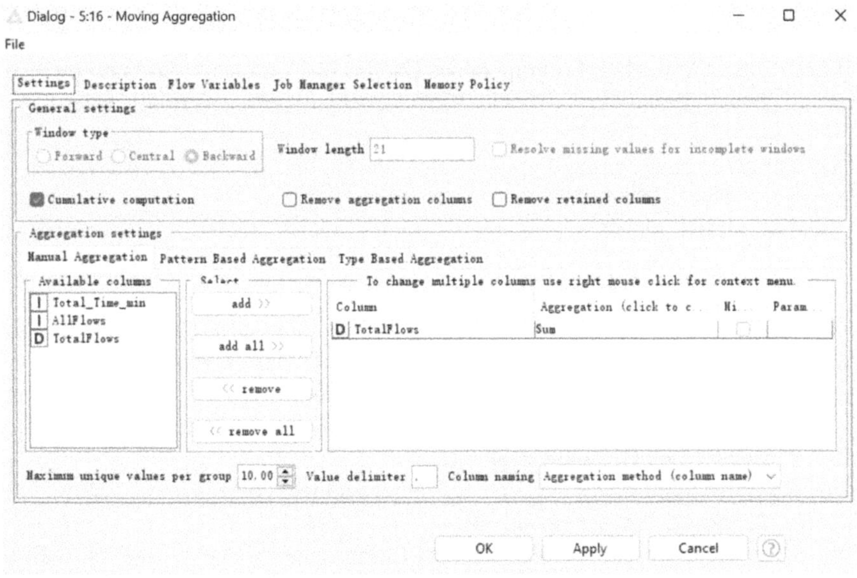

FIGURE 2.18   Interface of Moving Aggregation

Click OK and execute the node. This step adds a new column `Sum(TotalFlows)`, which calculates the cumulative percentage of flows ranging from 0 to 100.

Add a Math Formula node (Node 16), and link Moving Aggregation (Node 15) to it. Configure it to append a new column `ccdf` and calculate it as `100-$Sum(TotalFlows)$`. This step generates the complementary cumulative percentage of flows ranging from 100 to 0.

Step 5. Estimating the Distance Decay Functions for the Complementary Cumulative Distribution: To prepare for the regression analysis, we need to remove the long tail of the distribution. Add a Row Filter node (Node 17) and link Math Formula (Node 16) to it. In the dialog, keep `Include rows by attribute value` checked, choose `Total_Time_min` for Column to test, check `use range checking`, and input 180 for upper bound. Click OK and execute it.

Add a Line Plot node (Node 18) and link Row Filter (Node 17) to it. In the Line Plot dialog, set `Total_Time_min` for Choose column for x-axis, and select `Sum(TotalFlows)` and `ccdf` in the Include box for Choose column for y-axis. View the result as shown in Figure 2.19.

Similar to Step 1, add a Math Formula (Node 19), and link the outport data port of Line Plot (Node 18) or Row Filter (Node 17) to it. Configure it to append a new column `LnI` and calculate it as `ln($TotalFlows$)`.

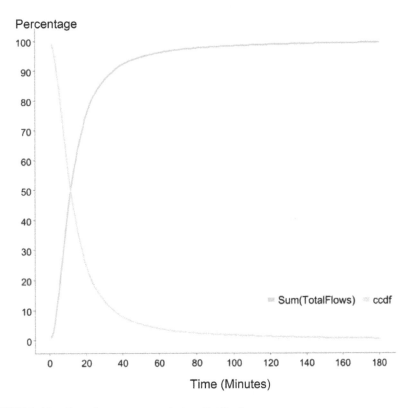

FIGURE 2.19 Complementary cumulative distribution curve

Since the data shares the same column names for regression in Step 1, we can use the same workflow consisting of Nodes 4–11. On the Editor Canvas, select Nodes 4–11 by clicking and pressing the Shift (Control) key, or dragging a box while pressing the left mouse button. Right-click one of the nodes and choose `Copy`. Right-click on an empty space on the Editor canvas and choose `Paste`. The nodes will be automatically renamed as Nodes 20–27. Move the nodes and links to a relatively close position to Math Formula (Node 19), and link Math Formula (Node 19) to the Math Formula (Node 20). Click the button ⊙ on the top bar (or press Shift+F7) to run all the nodes.

The best-fitting distance decay function is the log-normal function `lnI~Sq_LnD` with $R^2 = 0.9538$.

2.3   CONCLUDING REMARKS

This chapter uses several built-in nodes, including CSV Reader and CSV Writer in the IO category to read and write csv files and their zipped format. New Manipulation nodes include Concatenate for vertical data appending, String Manipulation for handling string columns and Round Double for output table rendering with varying levels of decimal precision.

For visualization, we use Line Plot introduced in Chapter 1 and a new node Scatter Plot. Under the Analytics category, newly introduced nodes include Linear Correlation and Linear Regression Learner for data comparison and OLS regression. Moving Aggregation for time-series data is used to calculate the Complementary Cumulative Distribution.

For Geospatial Analytics nodes, in addition to the nodes used in Chapter 1 such as GeoFile Reader, Geospatial View, Projection and Geometry to Point, two new nodes are introduced: Geocoding and Lat/Lon to Geometry, which are used in Spatial Conversion to convert tables to geospatial data in KNIME AP. The node Coordinates XYZ is used to extract the coordinates.

OSRM Distance Matrix, Google Distance Matrix, and Road Network Distance Matrix in the Spatial Network category are used to calculate travel distance and times based on a road network, and Euclidean Distance and Haversine Distance in the Spatial Manipulation category are simple distance measures based on spatial location. Access to Geocoding and Google Distance Matrix may require an API Key from the service providers.

NOTES

1 When dealing with a large dataset, modify the settings in the Advanced settings tab, e.g., uncheck Limit memory per column, or Limit data rows scanned.
2 https://raw.githubusercontent.com/UrbanGISer/CGA-KNIME-Workbook/main/data/chp2/BR_Bkg.zip
3 Haversine distance refers to the great-circle distance between two points on a sphere given their longitudes and latitudes.
4 https://developers.google.com/maps/documentation/distance-matrix/get-api-key
5 https://project-osrm.org/
6 In KNIME 4.7, the node name is Column Rename.

3 Spatial Smoothing and Spatial Interpolation

This chapter covers two generic tasks in GIS-based spatial analysis: spatial smoothing and spatial interpolation. Both are useful to visualize spatial patterns and highlight spatial trends. Spatial smoothing computes the average values of a variable in a larger spatial window to smooth its variability across space. Spatial interpolation uses known (observed) values at some locations to estimate (interpolate) unknown values at any given locations.

There are three case studies. The first case study of place names in southern China illustrates some basic spatial smoothing and interpolation methods. The second illustrates how to use area-based spatial interpolation methods to transform population data between different census areal units. The third demonstrates how to use the spatio-temporal kernel density estimation (STKDE) method for detecting spatiotemporal crime hotspots.

3.1   CASE STUDY 3A: MAPPING PLACE NAMES IN GUANGXI, CHINA

This case study examines the distribution pattern of contemporary Zhuang place names (toponyms) in Guangxi, China, based on a study reported in Wang et al. (2012). Zhuang, part of the Tai language family, are the largest ethnic minority in China and mostly live in the Guangxi Zhuang Autonomous Region (a provincial unit simply referred to as "Guangxi" here). The Sinification of ethnic minorities, such as the Zhuang, has been a long and ongoing historical process in China. The impact has been uneven in the preservation of Zhuang place names. The case study is chosen to demonstrate the benefit of using GIS in historical-linguistic-cultural studies. Spatial analysis techniques such as spatial smoothing and spatial interpolation methods can help enhance the visualization of the spatial pattern of Zhuang place names.

The data folder `Guangxi` includes:

1. point feature `Twnshp.zip` for all townships in the region with the field `Zhuang` identifying whether a place name is Zhuang (=1) or non-Zhuang (=0) (mostly Han), and
2. two polygon features `County.zip` and `Prov.zip` for the boundaries of counties and the provincial unit, respectively.

Figure 3.1 shows the workflow for this case study.

We begin by implementing the floating catchment area method. Different window sizes are experimented in order to identify an appropriate window size that balances overall trends and local variability.

DOI: 10.1201/9781003304357-3

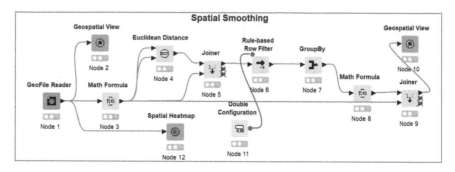

FIGURE 3.1   Workflow of Case Study 3A

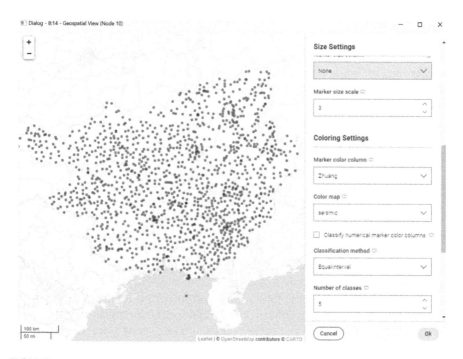

FIGURE 3.2   Mapping Zhuang and non-Zhuang place names in Guangxi, China

Step 1. Mapping Zhuang place names: Create a new workflow Case3A, add a GeoFile Reader node (Node 1) to the Editor, and input the file path of Twnshp.zip in the dialog, e.g., E:\KNIME\CMGISV3\Chp03\data\GX\Twnshp.zip. Click OK and execute it to read the data.

Add a Geospatial View node (Node 2) and link the GeoFile Reader (Node 1) to it. In the dialog shown in Figure 3.2, input 3 for Marker size scale, choose Zhuang for Marker color column, seismic for Color map, uncheck Classify numerical marker color column, and choose `CartoDB PositronNoLabels` for base map. Click OK and execute it.

Step 2. Computing distance matrix between places: Before performing the distance matrix calculation, we need to create an ID for the data joining that follows. Add a Math Formula node and link GeoFile Reader (Node 1) to it. Configure Math Formula (Node 3) to append a new integer column ID with the expression `$$ROWINDEX$$+1` (be sure to check `Convert to Int`).

Add a new node Euclidean Distance (Node 4) and link Math Formula (Node 3) to the two input ports. In the dialog of Euclidean Distance, input `ID` for both Origin ID Column and Destination ID column, keep `geometry` for both Origin geometry column and Destination geometry column, choose `Kilometer` for Distance unit. Click OK and execute it. The result table contains three columns, `Origin ID`, `Destination ID` and `Distance`, and the values in the first two columns are identical to the column `ID` of the input data.

Add a Joiner node (Node 5), link Euclidean Distance (Node 4) and Math Formula (Node 3) to it. In the dialog of Joiner, (1) under Joiner settings tab, choose `Destination ID` for Top Input ('left' table), `ID` for Bottom Input ('right' table); (2) under Column Selection, keep all columns in the Include box for both Top Input (left table) and Bottom Input (right table); and (3) keep other default settings. Execute it. This step attaches the attributes of points to the OD distance list.

Add a Rule-based Row Filter (Node 6) node to the Editor and link Joiner (Node 5) to it. In the dialog shown in Figure 3.3, input `$Distance$ <= 10 => TRUE`,

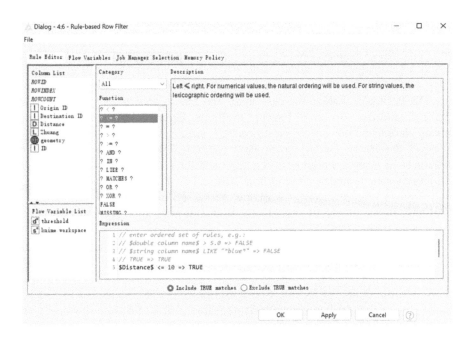

FIGURE 3.3 Using Rule-based Row Filter to define a catchment window

check `Include TRUE matches`. Click OK and execute it. This defines the window size with a radius of 10 km.

Step 3. Summarizing places within a window around each place: Add a new GroupBy node and link Rule-based Row Filter node to it. In the dialog of GroupBy (Node 7), (1) under the Groups tab, choose `Origin ID` for Group column(s) and `Aggregation method (column name)` for Column naming, (2) under manual Aggregation tab, choose `Zhuang` and `ID` for Column, and `Sum` and `Count` for Aggregation, respectively. Click OK and execute it. After execution, you will have two resulting columns: `Sum(Zhuang)` and `Count*(ID)`. The column `Sum(Zhuang)` represents the total number of Zhuang places, while the column `Count*(ID)` represents the total count of all places.

When using `Aggregation method (column name)` for column naming, there might be an asterisk * included in the names, such as `Count*ID`. This indicates that there are no missing values in that column. If you prefer to keep the name as `Count(ID)`, you can check the box under Missing located between the Aggregation settings and Parameters.

Step 4. Calculating and mapping Zhuang place ratios around each place: Add a Math Formula node (Node 8), and link GroupBy (Node 7) to it. Configure it to append a new column `ZhuangR` to measure the portion of Zhuang place names among all places within the window with the expression `$Sum(Zhuang)$/$Count*(ID)$`.

Add a Joiner node (Node 9), link Math Formula (Node 8) to the top input port of Joiner (Node 9), and link Math Formula (Node 3) to the bottom input port. In the dialog of Joiner (Node 9), (1) under Joiner settings tab, choose `Origin ID` for Top Input ('left' table), `ID` for Bottom Input ('right' table); (2) under Column Selection, choose `ZhuangR` in the Include box for Top Input (left table), choose all columns (`Zhuang`, `geometry`, `ID`) in the Include box for Bottom Input (right table), and keep other default settings. Execute it.

Add a Geospatial View node (Node 10) and link Joiner (Node 9) to it. In the dialog, choose `ZhuangR` for Marker size column and Marker color column, choose CartoDB PositronNoLabels for base map, and check `Classify numerical marker color column`. Execute the node. Figure 3.4 shows the smoothed Zhuang place name ratios within a 10-km FCA size.

Step 5. Sensitivity analysis with flow variable: This step demonstrates the advantages of using a workflow-based approach for data analysis. Use ⎚ Double Configuration node under ⚛ Workflow Abstract > ✖ Configuration > ▢ Input, or ⎚ Double Widget under ⚛ Workflow Abstract > ⊡ Widget > ▢ Input. Both nodes have the same function, but Double Widget is more appropriate for workflows designed for the KNIME Web portal.

Add a Double Configuration node to the Editor and link it to the left upper corner of Rule-based Row Filter (Node 5) as a flow variable. To show the flow variable ports, users can right click any node and choose `Show Flow Variable Ports`. In the dialog of Double Configuration (Node 11) shown in Figure 3.5, input `threshold` for Parameter/Variable Name, and 10 for Default value. Click

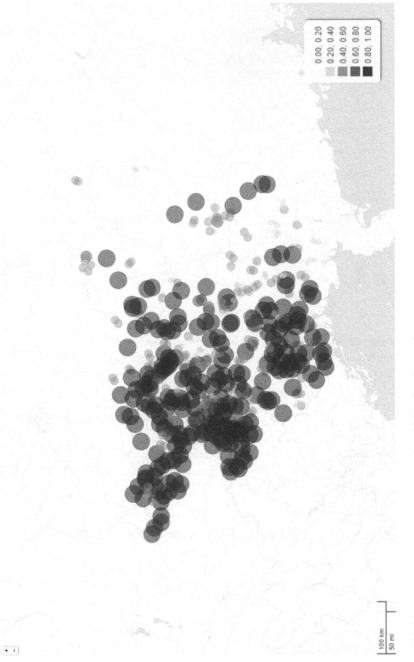

FIGURE 3.4 Zhuang place name ratios in a 10-km floating catchment area

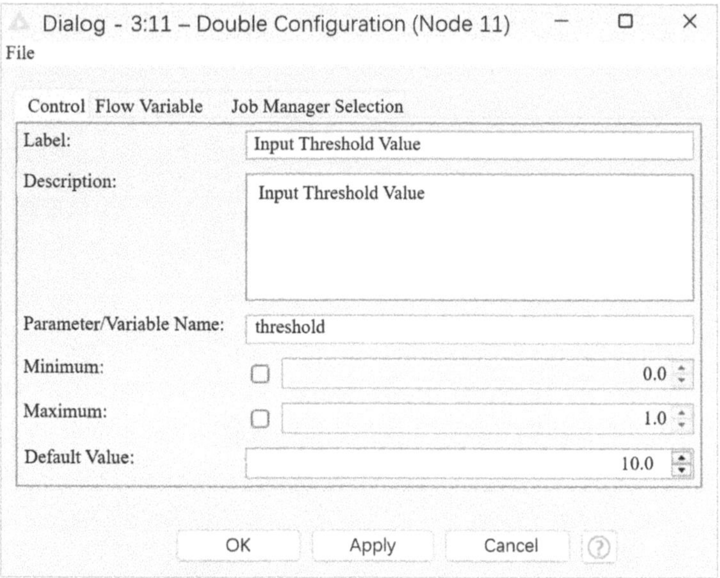

FIGURE 3.5   Interface of Double Configuration

OK and execute it. The node attaches the flow variable threshold with the value of 10 km. We can check the result by right clicking the node and choosing Flow Variable Output.

In the dialog of Rule-based Row Filter (Node 6), revise the expression as $Distance$ <= $${Dthreshold}$$ => TRUE. Re-execute the whole workflow. It implements the FCA method with a window size of 10 km which is represented by the flow variable threshold.

Try other window sizes such as 15 km by revising the value in Double Configuration (Node 11) and re-execute the workflow. This completes the implementation of FCA with various window sizes for spatial smoothing.

Step 6. Implementing the kernel density estimation method: Under 🌑 Spatial Visualization, drag ⬤ Spatial Heatmap to the Editor and link GeoFile Reader (Node 1) to it. In the dialog of Spatial Heatmap (Node 12), choose Zhuang for Weight column, choose CartoDB Positron for base map and keep the default settings for Minimum opacity, Maximum zoom, and Radius and Blur. Click OK and execute it. Right click the node and select Interactive View: Heatmap View. In the resulting map shown in Figure 3.6, click the white control box in the upper right to turn off the visualization layer.

This node creates an interactive heatmap. The selected weight column determines the "heat" of each data point. Users can experiment with various values for Radius parameter to see the effect of spatial smoothing on the visualization.

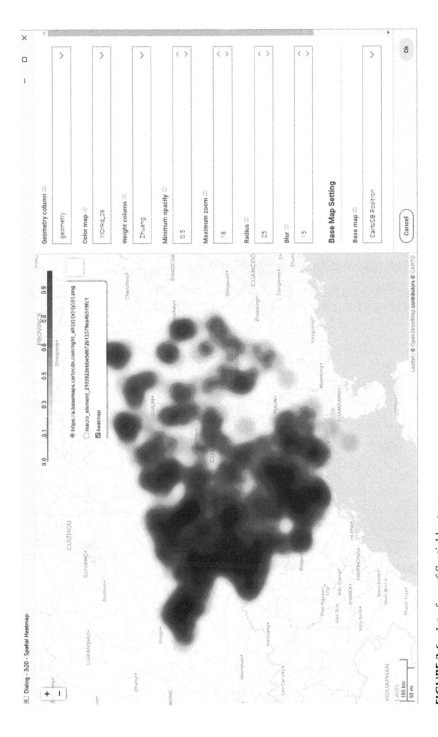

FIGURE 3.6 Interface of Spatial heatmap

3.2   CASE STUDY 3B: AREA-BASED INTERPOLATIONS OF POPULATION IN BATON ROUGE, LOUISIANA

Transforming data from one area unit to another or areal interpolation is a common task in spatial analysis for integrating data of different scales or resolutions. This case study illustrates two methods: areal weighting interpolation and Target-Density Weighting (TDW) Interpolation.

Data needed for the project are provided in the subfolder BR under the folder BatonRouge:

1. features BRTrt2010.zip and BRTrtUtm.zip for census tracts in the study area in 2010 and 2020, respectively, and
2. feature BRUnsd.zip for unified school districts in the study area in 2020.

Figure 3.7 shows the workflow for this case study.

3.2.1   IMPLEMENTING AREAL WEIGHTING INTERPOLATION

The Areal Weighting Interpolation method is illustrated in transforming population data from census tracts to school districts in the year 2020.

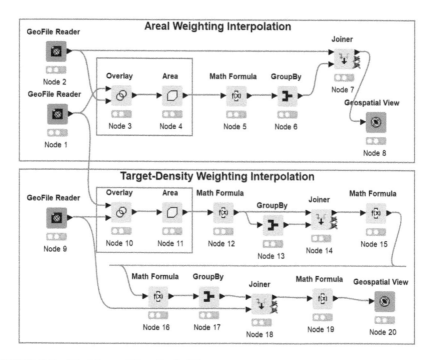

FIGURE 3.7   Workflow of Case Study 3B

Step 1. Overlaying census tract and school district layers: Create a new workflow
Case3B, and add two GeoFile Reader nodes to the Editor. Input the file paths for
BRTrtUtm and BRUnsd in the GeoFile Reader Nodes 1 and 2, respectively, e.g.,
E:\CMS_GISV3\Data\BR\BRTrtUtm.zip   and   E:\CMS_GISV3\Data\BR\
BRUnsd.zip.

Add a new node Overlay (Node 3) and link the two GeoFile Reader nodes to its
input ports. Be sure to put the node of BRTrtUtm as the first input port of Overlay
(Node 3) as the CRS information of the first input port will be applied to all input
features. In the dialog of Overlay (Node 3), choose Intersection for Overlay
mode, and check Return only geometries of the same geometry
type. Execute the node.

Add a new node Area (Node 4) and link Overlay (Node 3) to it. Execute it to attach
the area of intersection units with a new column area.

Step 2. Apportioning attribute to area size in intersected polygons: Add a new
node Math Formula (Node 5) and link Area (Node 4) to it. Configure it to append a
new column EstPopu with the expression $POP100$/$Area$* $area$. This
is the interpolated population for each polygon in the intersected layer (atom) by the
areal weighting method.

Step 3. Aggregating data to school districts: Add a new GroupBy node (Node
6) and link Math Formula (Node 5) to it. In the dialog, (1) under Groups tab, add
DISTRICT_N in the box of Group Column(s), choose Keep original name(s)
for Column naming; (2) under Manual Aggregation, choose EstPopu for Column
and Sum for Aggregation. Execute the node. The interpolated population for school
districts is in the column EstPopu in the resulting table.

Add a Joiner node (Node 7) and link GeoFile Reader (Node 2) and GroupBy
(Node 6) to its input ports. In the dialog of Joiner (Node 7), under Joiner settings tab,
choose DISTRICT_N for both Top Input ('left' table) and Bottom Input ('right' table),
and keep the default setting of Column Selection to include all columns. Execute it.
Hereafter, we will simply rephrase it as "Join the two tables based on the common
column DISTRICT_N and include all columns."

Add a Geospatial View node (Node 8) to map the values of EstPopu as shown
in Figure 3.8.

3.2.2   IMPLEMENTING TARGET-DENSITY WEIGHTING (TDW) INTERPOLATION

It is common that census tract boundaries change from one decennial year (2010) to
another (2020). In order to map and analyze the population change over time, one
has to convert the population in two different years to the same areal unit. The task
here is to interpolate the 2020 population in the 2020 census tracts to the 2010 census
tracts by the TDW method so that population change rates can be assessed in the
2010 census tracts.

According to the TDW, the layers of 2010 and 2020 census tracts are intersected
to create a layer of atoms, and the ancillary variable (z or 2010 population) in the
target zone (t or 2010 census tracts) is firstly distributed (disaggregated) across the

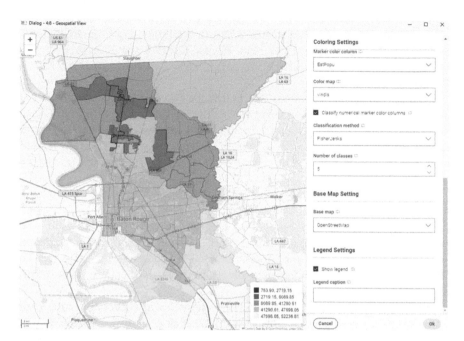

FIGURE 3.8   Population across school districts in Baton Rouge

atoms (*st*) assuming uniformity, and then aggregated in the source zone (*s*, i.e., 2020

census tracts), producing the proportion $\dfrac{\hat{Z}_{st}}{\hat{Z}_s}$. The variable of interest (*y* or 2020

population) in the source zone is subsequently disaggregated to atoms according to this proportion to honor the spatial distribution pattern of the ancillary variable (*z* or 2010 population). The disaggregated *y* (2020 population) is then aggregated in the target zone (2010 census tracts) to complete the interpolation process. In summary, the TDW re-distributes the population of 2020 to match the 2010 census tract boundary in a manner to honor the historical distribution pattern of the population of 2010. In other words, when interpolating the 2020 population to the atoms, the TDW method assumes that it is more reliable to allocate it proportionally to the historical (2010) population pattern (the ancillary variable) than proportionally to the area sizes (i.e., the areal weighting approach).[1] Such an assumption was validated by Schroeder (2007).

Step 4. Overlaying the two census tract layers: Similar to step 1, add a GeoFile Reader node (Node 9) and input the file path of BRTrt2010 in it, e.g., E:\CMS_ GISV3\Data\BR\BRTrt2010.zip. Copy and paste the two nodes Overlay and Area (Nodes 3 and 4) and their link as two new nodes (Nodes 10 and 11), and link GeoFile Reader (Node 1) and GeoFile Reader (Node 9) to the input ports of Overlay (Node 10). Execute the two nodes with the predefined parameters.

Step 5. Disaggregating 2010 population to the atoms. Add a Math Formula node (Node 12) and link Area (Node 11) to it. Configure the node to append a new column EstP2010 with the expression $DP0010001$/ $Area_2$* $area$, where area is the area size of atoms and Area_2 is the area size of 2010 tracts. Similar to step 2, this proportions the 2010 population to the areas of atoms. The field EstP2010 is $\hat{Z}_{st}$.

Step 6. Aggregating estimated 2010 population to 2020 tracts: Add a GroupBy node (Node 13) and link Math Formula (Node 12) to it. In the dialog, (1) under Groups tab, input GEOID for Group column(s), and choose Aggregation method (column name) for Column naming; (2) under Manual Aggregation, choose EstP2010 for Column and Sum for Aggregation. Execute the node. The column Sum(EstP2010) in the resulting table is $\hat{Z}_s$.

Step 7. Computing the ratio of 2010 population in atoms out of 2020 tracts: Add a Joiner node (Node 14), link Math Formula (Node 12) and GroupBy (Node 13) to its input ports. Join the two tables based on the common column GEOID and include all columns.

Add a Math Formula node (Node 15) and link Joiner (Node 14) to it. Configure it to append a new column P10_rat with the expression $EstP2010$/ $Sum(EstP2010)$, which is the proportion $\dfrac{\hat{Z}_{st}}{\hat{Z}_s}$.

Step 8. Disaggregating 2020 population to atoms according to the ratio: Add a Math Formula node (Node 16) and link Math Formula (Node 15) to it. Con-figure it to append a new column EstP2020 with the expression $POP100$*$P10_rat$. In other words, the TDW method interpolates 2020 population in atoms proportionally to 2010 population instead of area sizes as the areal interpolator does.

Step 9. Aggregating estimated 2020 population to 2010 tracts: Add a GroupBy node (Node 17) and link Math Formula (Node 16) to it. In the dialog, (1) under Groups tab, input GEOID10 for Group column(s), and choose Aggregation method (column name) for Column naming; (2) under Manual Aggregation, choose EstP2020 for Column and Sum for Aggregation. Execute the node. The field Sum(EstP2020) in the result table is the interpolated 2020 population in the 2010 tracts.

Step 10. Mapping population change rates from 2010 to 2020 in 2010 tracts: Add a Joiner node (Node 18), and link GroupBy (Node 17) and GeoFile Reader (Node 9) to it. Join the two tables based on the common column GEOID10 and include all columns.

Add a Math Formula (Node 19) and link Joiner (Node 18) to it. Configure it to append a new column ChangeR with the expression ($Sum(EstP2020)$- $DP0010001$)/$DP0010001$.

Add a new node Geospatial View (Node 20) to map the field ChangeR (population change rate) as shown in Figure 3.9.

FIGURE 3.9 Population change rates across census tracts in Baton Rouge 2010–2020

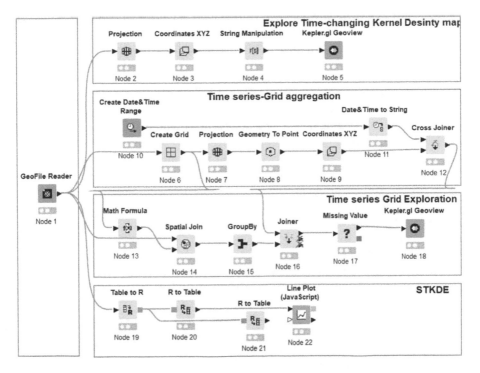

FIGURE 3.10   Workflow of Case Study 3C

3.3   CASE STUDY 3C: DETECTING SPATIOTEMPORAL CRIME HOTSPOTS IN BATON ROUGE, LOUISIANA

This case study explores the spatiotemporal patterns of residential burglary crimes in Baton Rouge, Louisiana in 2011, as reported in Hu et al. (2018).

Data needed for the project is provided in a sub-folder BRcrime under the folder BatonRouge. It contains a zipped Shapefile BRcrime.zip for residential burglary crimes with XY coordinates and time label, and an R file stkde.r for implementing the STKDE model.

The following steps illustrate how to detect spatial and temporal clusters, create a space time analysis frame and analyze emerging hotspots with the STKDE. Figure 3.10 shows the workflow for this case study.

3.3.1   EXPLORING TIME SERIES HEATMAP BASED ON KEPLER.GL GEOVIEW

Step 1. Preparing data for spatial temporal change visualization with Kepler.gl GeoView: Create a new workflow Case3C. Add a GeoFile Reader node (Node 1) to the Editor and input the file path for BRcrime, e.g., E:\KNIME\CMGISV3\Chp03\data\BRcrime.zip. The resulting table contains five columns: ID, X, Y, time, and geometry, with a CRS of EPSG:26915.

As point visualization in the Kepler.gl GeoView node requires geographic instead of projected coordinates, a CRS transformation is applied first. Add a Projection node

(Node 2) and link GeoFile Reader (Node 1) to it. In the dialog of Projection (Node 2), set `epsg:4326` as the New CRS, check `Replace` for Output column and execute it. This transforms the geometry to the WGS 84 geographic coordinate system.

Add a Coordinates XYZ node (Node 3), and link Projection (Node 2) to it. Execute it and observe the difference between the columns of `x` and `X`, `y` and `Y`. Now the columns `x` and `y` represent longitude and latitude, respectively.

The time column needs to be converted to the format of `YYYY-MM-DDThh:mm:ss` (e.g., 2011–10–01T00:00:00) for using time-based map animation in Kepler.gl GeoView. Add a String Manipulation node (Node 4) to the workflow. In the dialog, input `join($time$,"T00:00:00")` in the expression field, check `Append Column`, input `DataTime` as the column name, and keep other default settings. Click OK and execute the node. This converts time to the new format.

Step 2. Spatiotemporal change visualization with Kepler.gl GeoView: Add a Kepler.gl Geoview node (Node 5) and link String Manipulation (Node 4) to it. Execute it. Right click on Kepler.gl Geoview (Node 5) and choose "Interactive View: Geospatial View" to open the visualization window.

Completing the Geospatial View dialog consists of multiple steps. As shown in Figure 3.11a, (1) click the hidden menu button In the upper left corner ⟩ to expand

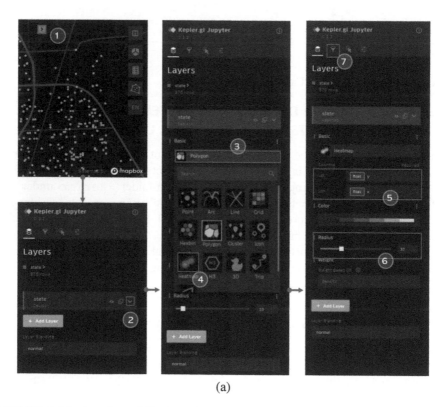

(a)

FIGURE 3.11   Interface of Kepler.gl Geoview for time series heat map: (a) items 1–6, (b) items 7–9

(Continued)

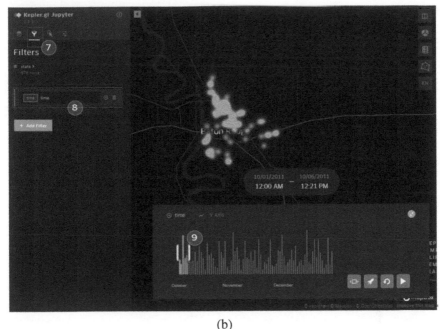

(b)

FIGURE 3.11 Continued

the menu; (2) click the drop-down menu ∨ on the `state` block to expand the layer menu; (3) click Polygon under Basic to bring out the visualization geometry type menu; (4) click Heatmap; (5) and choose y for Lat, x for Lon to view the map as a heatmap based on kernel density estimation; adjust the slider to increase the bandwidth; and (6) set the Radius to 30. Continue as shown in Figure 3.11b, (7) click the button ♥ to switch to the Filter tab; (8) click Add filter and choose `time` for it; and (9) on the lower right pop-up control panel, adjust the time window size, animation speed, and other parameters as desired.

3.3.2 IMPLEMENTING TIME SERIES HEATMAP BASED ON GRID AGGREGATION

Step 3. Creating 1km × 1km grid: Under Spatial Manipulation, drag Create Grid node to the Editor, and link GeoFile Reader (Node 1) to it. In the dialog of Create Grid (Node 6), set the Grid Length to 1000 to create a 1km × 1km grid with an ID column `Grid ID` based on the extent and CRS of the input data.

Add three new nodes, Projection, Geometry To Point, and Coordinate XYZ. Link the nodes in the order of Create Grid (Node 6) → Projection (Node 7) → Geometry To Point (Node 8) → Coordinate XYZ (Node 9). Use the default settings for each node to transform the grid into its centroids in the geographic coordinate system, with new coordinates, x and y.

Step 4. Creating a full Date & Time range: Under ⚙ Other Data Type > ∿ Time Series > ⬜ Manipulation, drag Create Date&Time Range (Node 10). In the dialog

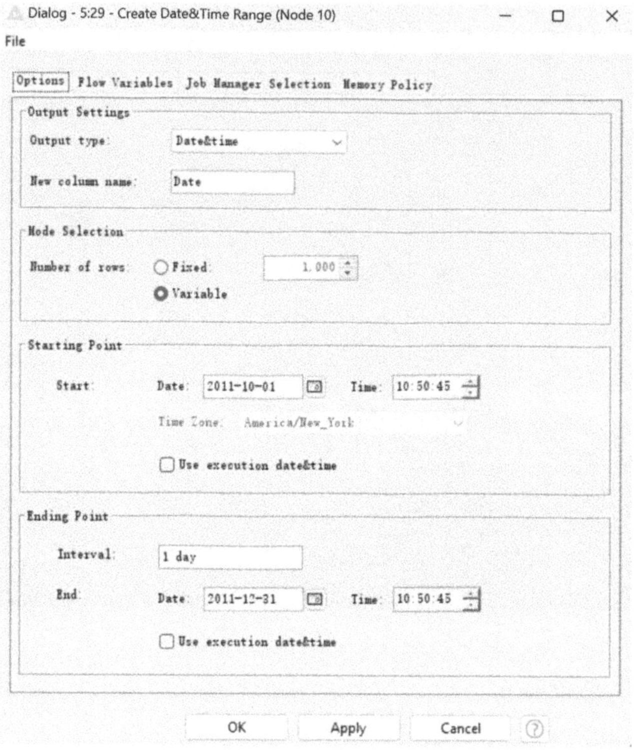

FIGURE 3.12 Interface of Create Date&Time Range

shown in Figure 3.12, (1) under Output Settings, choose Date&time for Output type and Date for New column name; (2) under Mode Selection, choose Variable for Number of rows; (3) under Starting Point, input 2011-10-01 for Date, and keep the default value for Time; (4) under Ending Point, input 1 day for Interval and 2011-12-31 for Date. Click OK and execute the node. The result table contains only one column Date in the format of YYYY-MM-DDThh:mm:ss.

Under ⁰ Other Data Type > ～ Time Series > ☐ Transform, drag Date&Time to String to the Editor, and link Create Date&Time Range (Node 10) to it. In the dialog of Date&Time to String (Node 11) shown in Figure 3.13, choose Date in the Include box, check Append selected column, keep the default value (string) for Suffix of appended columns, choose yyyy-MM-dd for Date format, and keep other default settings. Click OK and execute it. This node appends a new column Date (string) to the input table in the predefined format.

Add a Cross Joiner node (Node 12), and link Date&Time to String (Node 11) and Coordinate XYZ (Node 9) to its two input ports. Execute it. The whole workflow from Node 6 to Node 12 (Steps 3–4) builds a 3-dimensional grid with the key columns (x, y and Date) based on the 2D grid.

FIGURE 3.13   Interface of Date&Time to String

Step 5. Aggregating crime data to the 3D grid: Add a Math Formula node (Node 13) and link Create Grid (Node 6) to it. Configure it to append a new integer column count with the expression 1. This appends a column populated with value 1.

Add a Spatial Join node (Node 14), and link GeoFile Reader (Node 1) and Math Formula (Node 13) to it. In the dialog, choose Inner for Join mode and Intersects for Match mode. The node appends the attributes of each 2D grid to the points that intersect with it.

Add a GroupBy node (Node 15) and link Spatial Join (Node 14) to it. In the dialog of GroupBy (Node 15), (1) under the Groups tab, choose time and Grid ID for Group column(s), Keep original name(s) for Column naming; (2) under Manual Aggregation tab, choose count for Column and Sum for Aggregation. Click OK and execute it. The three columns time, Grid ID and count indicate the aggregated crime records in the 3D grid.

Step 6. Joining the aggregated data to the full 3D grids: Add a Joiner node (Node 16), link Cross joiner (Node 12) to its top input port, and link GroupBy (Node 15) to its bottom port. In the dialog of Joiner (Node 16), (1) under Joiner settings tab, set Grid ID and Date(String) for Top Input ('left' table) and Grid ID and time for Bottom Input ('right' table); (2) under Include in outputs, check both Matching rows and Left unmatched rows, which can be seen as a left join; (3) under Column Selection tab, choose all columns in the Include box for Top Input (left table), choose count in the Include box for Bottom Input (right table), and keep other default settings. Execute it. The result table shows the aggregated crime records

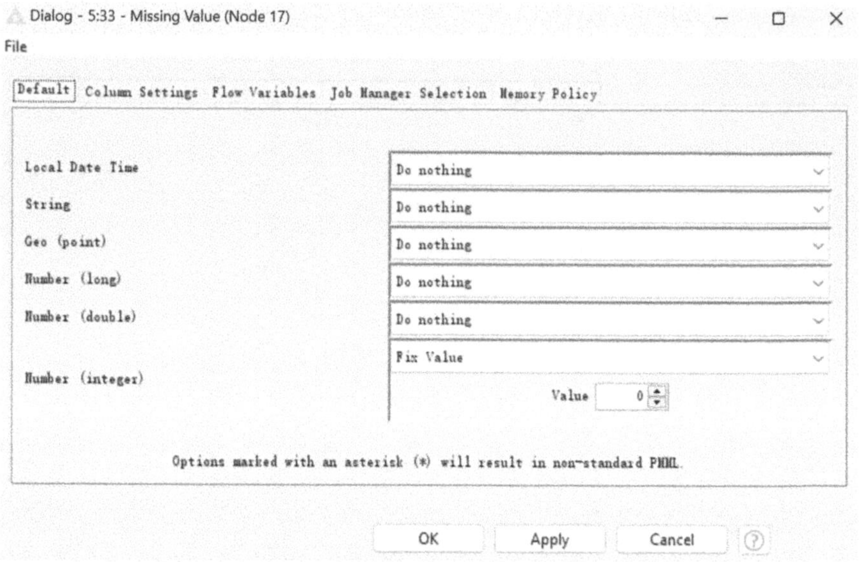

Dialog - 5:33 - Missing Value (Node 17) — □ ✕

File

| Default | Column Settings Flow Variables Job Manager Selection Memory Policy |

Local Date Time	Do nothing ⌄
String	Do nothing ⌄
Geo (point)	Do nothing ⌄
Number (long)	Do nothing ⌄
Number (double)	Do nothing ⌄
Number (integer)	Fix Value ⌄
	Value 0 ⬍

Options marked with an asterisk (*) will result in non-standard PMML.

OK   Apply   Cancel   ⑦

FIGURE 3.14   Interface of Missing Value

in the 3D grid, with missing values in the count column for days or grids with no crime records.

Under Manipulation > Column > Transform, drag a Missing Value node (Node 17) to the Editor and link Joiner (Node 16) to it. In the dialog shown in Figure 3.14, choose Fix Value for Number(integer) and input 0 for Value, and keep other default settings. Click OK and execute it. All the missing values denoted by "?" in the integer column count are replaced with 0, indicating no crime records on those days or grids.

Step 7. Visualizing the time-series grid data in Kepler.gl GeoView by Grid, Hexibin, Point and Heatmap: Add a Kepler.gl Geoview node (Node 18) and link Missing Value (Node 17) to it. Execute it directly. Right click on Kepler.gl Geoview (Node 18) and choose "Interactive View: Geospatial View" to open the visualization window.

In the window (refer to Figure 3.11), (1) click the hidden menu button ❬ to expand the menu; (2) click the drop down menu ⌄ on the state to expand the layer menu; (3) click Polygon under Basic to bring out the visualization type menu (Figure 3.15a); (4) click Point, and choose y for Lat, x for Lon; (5) click the function icon ⁝ of Fill Color to expand the fold menu, choose count for Color Based On and quantize for Color Scale; (6) click ▼ Filters to switch to the Filter tab, click Add filter and choose Date for it. Adjust the time window size on the lower right pop-up control panel to view the spatiotemporal trend.

Users can choose different visualization options for Grid, Hexbin and Heatmap as illustrated in Figure 3.15b–d. For example, the Grid view with parameters set as in

Figure 3.15b displays the Grid map with height in 3D map by clicking the icon ![icon], or changes the base map by clicking ![icon] and adjusting the parameters in the base map pane.

Figure 3.16 shows the corresponding visualization results based on the parameter settings in Figure 3.15.

In the Kepler.gl Geoview dialog window (refer to Figure 3.11b), click Add filter and choose `time` for it, then on the lower right pop-up control panel, adjust the time window size, animation speed and other parameters to view its temporal patterns and trends.

3.3.3 IMPLEMENTING SPATIOTEMPORAL KERNEL DENSITY ESTIMATION

This subsection illustrates how to use the R snippet node to apply the Spatio-Temporal Kernel Density Estimation (STKDE) for predictive hotspot evaluation.

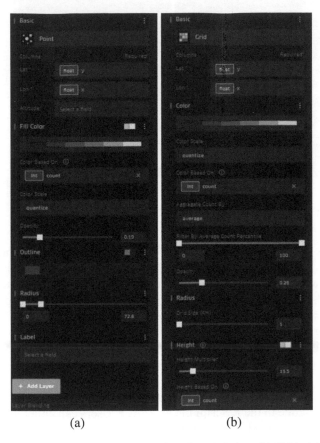

(a)   (b)

FIGURE 3.15   Interfaces of Kepler.gl GeoView for (a) Point, (b) Grid, (c) Hexbin, and (d) Heatmap

(Continued)

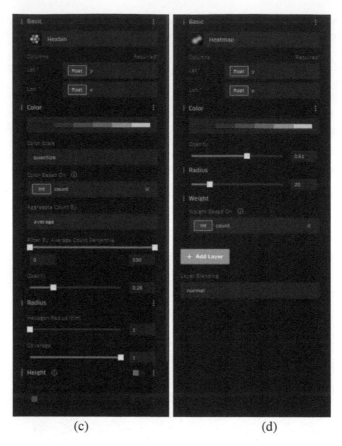

(c)   (d)

FIGURE 3.15   Continued

Note that the Table to R node can only accept one input table port and is not configurable like the Python Script node. One may need to use multiple "Add Table to R" nodes to add more tables. However, if only one tabular output dataset needs to be generated, use the R Snippet node with one pair of input and output ports for data.

Step 8. Applying STKDE *for predictive hotspot evaluation*: Under ⟷ Scripting > ʀ R, drag 🈁 Table to R to the Editor and link GeoFile Reader (Node 1) to Table to R (Node 19). The dialog (Figure 3.17a) contains Column list, Flow Variable List, R Script, Workspace, and Console. Double click the variable name in Flow Variable List to input that variable in R Script, e.g., double click `knime.workspace` to input `knime.flow.in[["knime.workspace"]]`.

Input the R code in the R Script file `stkde.r`. Users can either use Eval Script to run all the code or select the code line by line and click Eval Selection to test the code. While running the code in the dialog, the table of Workspace shows all the results, with related information in the Console. Click OK and execute the node. This node runs the STKDE model and generates two result tables,

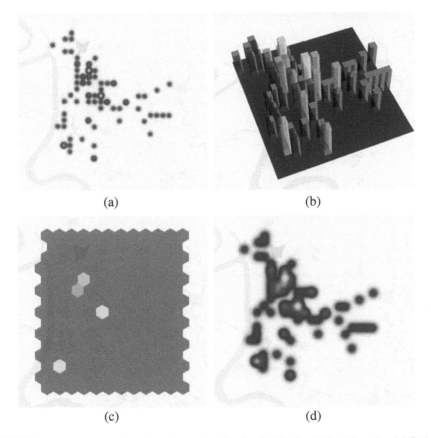

(a)　　　　　　　　　　　　(b)

(c)　　　　　　　　　　　　(d)

FIGURE 3.16 Time series crime data visualization in Kepler.gl GeoView for (a) Point, (b) Grid, (c) Hexbin, and (d) Heatmap

eva_result and df_result in the R workspace. To export the two data from R workspace, under ⟼ Scripting > ℝ R, drag two ℝ R to Table nodes (Node 20 and Node 21) to the Editor and link Table to R (Node 19) to them. In the dialog of R to Table (Node 20) as shown in Figure 3.17b, input knime.out <-eva_result in the R Script window; in the dialog of R to Table (Node 21) (not shown here), input knime.out <-df_result in the R Script window. The result table is exported via the port.

The result table of R to Table (Node 20) contains five columns: arearate, hirate, pai, nni and nni_p. They correspond to metrics commonly used to evaluate the predictive accuracy of identified crime hotspots: (1) *area rate* such as *a/A*, where *a* is area of identified hotspots from Time 1 (prior, say, the whole month of October 2011) data and *A* is the study area size; (2) *hit rate* such as *n/N*, where *n* is the number of robberies inside predicted hotspots and *N* is the total number of reported

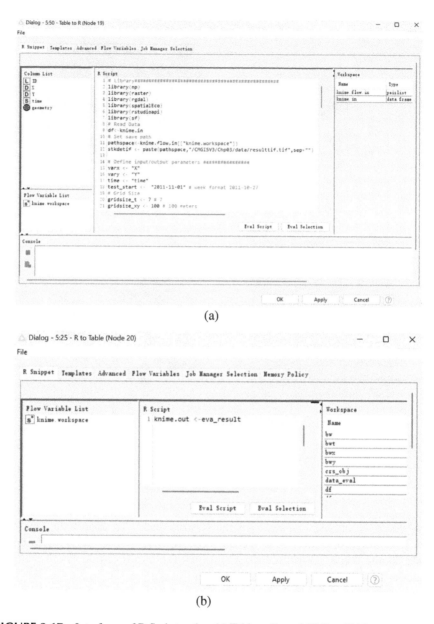

(a)

(b)

FIGURE 3.17   Interfaces of R Script nodes: (a) Table to R, and (b) R to Table

robberies in Time 2 (newer, say, the first week of November 2011); (3) *prediction Accuracy Index (PAI)* such as $PAI = \left(\dfrac{n}{N}\right) / \left(\dfrac{a}{A}\right)$, i.e., the ratio of the hit rate (n/N) to the area rate; (4) *nearest neighbor index (NNI)*, also known as Nearest neighbor distances, which computes observed distances between each point on a map and its nearest neighbor and (5) nni_P for the *p*-value of NNI based on simulation. See Hu

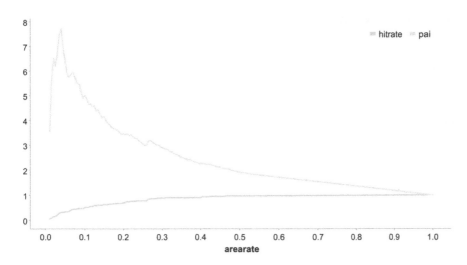

FIGURE 3.18   Line plots of hit rate and PAI vs. area rate

et al. (2018) for more details. The result table of R to Table (Node 21) shows predicted values from 2011–11–01 to 2011–11–07.

Users can add a Line Plot (Node 22) to view the result in Node 20. As shown in Figure 3.18, the hit rate increases with area rate, and the PAI peaks when area rate reaches about 0.35. One can also use Kepler.gl Geoview to map the predicted values in the result table of R to Table (Node 21) after applying projection transformation.

3.4   CONCLUDING REMARKS

The three case studies in this chapter implement various spatial smoothing and spatial interpolation techniques in KNIME AP. Users have another chance to practice several built-in nodes used in Chapters 1 and 2, such as CSV Reader, String Manipulation, Math Formula, Joiner, Cross Joiner, GroupBy, and Line Plot. Several new nodes in the Scripting and Workflow Abstract categories are also introduced.

In the Manipulation category, the Rule-based Row Filter node is more flexible than the Row Filter node, and the Missing Value node assigns default values for missing data to various data types.

The nodes of Create Date&Time Range and Date&Time to String are utilized to create and transform date and time with string format. The Script nodes of Table to R and R to Table enhance the modeling functionality of KNIME AP based on R Furthermore, the Double Configuration node in the Workflow Abstract category aids in workflow automation and iteration.

Some Geospatial Analytics Extension nodes used in the previous case studies, such as Spatial IO (GeoFile Reader), Spatial Visualization (Geospatial View), Spatial Manipulation (Euclidean Distance, Spatial Join, Projection, Geometry To Point, Overlay), Spatial Transformation (Projection, Geometry to Point), and Spatial Calculation (Coordinates XYZ, Area) are practiced again. In addition, the Create

Grid node is introduced to implement aggregation with standard spatial units, and the Spatial Heatmap node is used to visualize the spatial smoothing result. Kepler. gl Geoview is a powerful visualization tool based on methods such as Point, Grid, Hexbin, and Heatmap. In addition, it facilitates the exploration of time series animation by utilizing the specific time format tag YYYY-MM-DDThh:mm:ss.

NOTE

1   One may also transform the 2010 population data to the 2020 tracts and assess the changes in the 2020 tracts. In that case, the ancillary variable is the 2020 population.

4 Delineating Functional Regions and Application in Health Geography

A *functional region* is an area around a node, facility, or hub connected by a certain function (e.g., retail distribution, health care service, advertisement for a media company, or telecom coverage). By defining a functional region, a distinctive market area has a geographic boundary that encompasses many smaller areas more closely connected than beyond.

Defining functional regions is a classic task in geography. It can be as straightforward as the proximal area method that assigns areas to their nearest facility, or the Huff model that accounts for the joint effects of facility sizes and their distances from residents. Case Study 4A illustrates how to use the proximal area method and the Huff model to estimate the service areas of acute hospitals in Baton Rouge. Case Study 4B implements the Dartmouth method that pioneered the delineation of hospital service areas (HSAs) by a simple plurality rule, and the network community detection approach to define HSAs with maximal patient flows within HSAs and minimal flows between HSAs. It relies on data of observed hospitalization service volumes between residents and hospitals in Florida.

4.1 CASE STUDY 4A: DEFINING SERVICE AREAS OF ACUTE HOSPITALS IN BATON ROUGE, LOUISIANA

This case study defines the service areas of five acute hospitals in Baton Rouge based on travel time through the road network derived from Chapter 2. In the Huff model, we use the number of beds to define the capacity of each hospital S and assume the traditional power function for the distance decay effect with $\beta = 3.0$.

Data sets needed for the project are the same as in Case Study 2 under the data folder BatonRouge. Several data sets generated from Case Study 2 are also used. The data used includes:

(1) a projected census block group layer BR_Bkg.zip used in Chapter 2 and its centroids layer BRPt.zip generated in Step 8 of Subsection 2.1.2,
(2) a hospitals layer Hosp.zip generated in Step 8 of Subsection 2.1.2,
(3) a major road network dataset BR_MainRd.zip, and
(4) an OD drive time matrix file OD_Drive_Time.csv generated by Step 12 in Section 2.1 (its field Duration representing drive time in minutes and Distance for drive distance in meters).

DOI: 10.1201/9781003304357-4

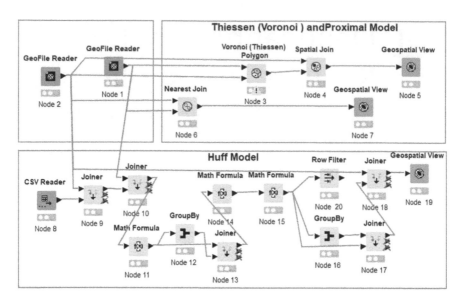

FIGURE 4.1 Workflow of Case Study 4A

Figure 4.1 shows the workflow for this case study.

4.1.1 Defining HSAs by the Proximal Area Method

This subsection describes two ways to define proximal areas by the supply-based and demand-based approaches, respectively.

Step 1. Defining Hospital Service Areas with Thiessen Polygons: Create a new workflow Case4A. Add two GeoFile Reader nodes to the Editor and input the file path of hospitals in Node 1 (e.g., E:\KNIME\CMGISV3\Chp04\data\Hosp.zip) and the file path of census block groups in Node 2 (e.g., E:\KNIME\CMGISV3\Chp04\data\BR_Bkg.zip), and execute them to read the data.

Under Spatial Manipulation, drag ⊕ Voronoi (Thiessen) Polygon (Node 3) to the Editor, link GeoFile Reader (Node 1 for hospitals) to the top input port, and link GeoFile Reader (Node 2 for census block groups) to the bottom input port. In the dialog shown in Figure 4.2a, keep 1000 for Buffer distance, Kilometer for Distance unit, and check Keep CRS from input table. Click OK and Execute it. The control buffer distance (in kilometers) is used to control the output of the polygons. If the final output polygon is smaller than the reference boundary, the buffer distance may need to be increased. This creates Thiessen polygons based on the input point data Hosp and reference boundary BR_Bkg (Figure 4.3a).

Add a Spatial Join node (Node 4), link GeoFile Reader (Node 2) to the top input port of it, and link Voronoi (Thiessen) Polygon (Node 3) to the bottom input port. In the dialog, keep Inner for Join mode, and choose Has its center in for Match mode. Click OK and execute it. This step allows all the census block group units inside each Voronoi polygon to inherit its attributes, e.g., hospital index Thiessen ID.

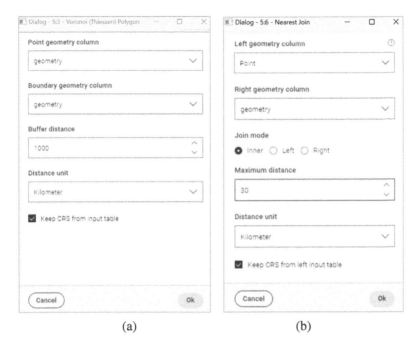

(a)   (b)

FIGURE 4.2 Interface of (a) Voronoi (Thiessen) Polygon and (b) Nearest Join

Add a Geospatial View (Node 5) and link Spatial Join (Node 4) to it. In the dialog, choose `Thiessen ID` for Marker color column and `CartoDB PositronNoLabels` for base map. Execute the node to view the result. This completes the delineation of hospital proximal areas by the supply-based approach. The result is shown in Figure 4.3b.

Step 2. Defining Hospital Service Areas with the Near tool: Before implementing the nearest joining, we need to generate centroids for all census block group units. Add a Geometry To Point node (Node 6) to the Editor and link GeoFile Reader (Node 2) to it. In the dialog of Node 6, check `centroid` for Point Type Selection and `Append` for Output column, click OK and execute it to append a point geometry column to the input data.

Add a Nearest Join node (Node 7) to the workflow and link Geometry To Point node (Node 6) to its top input port and GeoFile Reader (Node 1) to its bottom port. In the dialog shown in Figure 4.2b, Choose `Point` for Left geometry column keep `Inner` for Join mode, input 30 for Maximum distance and `Kilometer` for Distance unit, and check `Keep CRS from input table`. Click OK and execute the node. The joined column `OID_` in the result table indicates the nearest hospital from each census block group.

Link Nearest Join (Node 7) to Geospatial View (Node 5). In the dialog, choose `OID_` for Marker color column and execute it to view the result. This completes the delineation of hospital proximal areas by the demand-based approach. The result is shown in Figure 4.3c. It is largely consistent with Figure 4.3b.

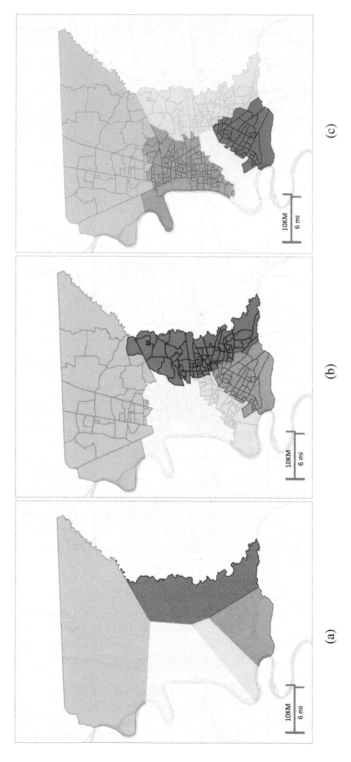

FIGURE 4.3   Proximal areas for hospitals: (a) Thiessen polygons, (b) supply-based approach, and (c) demand-based approach

4.1.2 DEFINING HSAS BY THE HUFF MODEL

Recall the Huff model written as, $P_{ij} = S_j d_{ij}^{-\beta} / \sum_{k=1}^{n} \left(S_k d_{ik}^{-\beta} \right)$, where S is a store's size, d is the distance or travel time, $\beta > 0$ is the distance friction coefficient, P_{ij} is the probability of an individual i selecting a store j, and k are the alternatives available ($k = 1, 2, \ldots, n$).

Step 3. Measuring hospital potentials: Add CSV Reader (Node 8) to the Editor and input the file path of OD Matrix, e.g., E:\KNIME\CMGISV3\Chp04\data\ OD_Drive_Time.csv. Execute it to read the data.

Add Joiner to the Editor, and link GeoFile Reader (Node 2) and CSV Reader (Node 8) to the two input ports of Joiner (Node 9). In the dialog of Joiner (Node 9), (1) under Joiner settings tab, choose OBJECTID for Top Input ('left' table) and Origin ID for Bottom Input ('right' table), and check making integer types compatible; (2) under Column Selection, choose Popu2020 in the Include box for Top Input (left table), choose all columns in the Include box for Bottom Input (right table), and keep other default settings. Execute it.

Link GeoFile Reader (Node 1) and Joiner (Node 9) to Joiner (Node 10). In the dialog of Joiner (Node 10), (1) under Joiner settings tab, choose OID_ for Top Input ('left' table) and Destination ID for Bottom Input ('right' table), and check making integer types compatible; (2) under Column Selection, choose Bed_size in the Include box for Top Input (left table), choose all columns in the Include box for Bottom Input (right table), and keep other default settings. Execute it.

The two Joiner nodes attach hospital service capacity and population to the OD matrix table. Add a Math Formula node (Node 11) and link Joiner (Node 10) to it. Configure it to append a new column Potent with the expression Bed_size* $Distance$^(-3). This computes the potential term $S_j d_{ij}^{-\beta}$ in the Huff model.

Step 4. Calibrating Total Hospital Potential for Each Census Block Group: Add a GrougBy node (Node 12) and link Math Formula (Node 11) to it. In the dialog, (1) under the Groups tab, choose Origin ID for Group column(s) and Aggregation method (column name) for Column naming; (2) under manual Aggregation tab, choose Potent for Column and Sum for Aggregation. Click OK and execute it.

In the resulting table, the column Sum(Potent) is the total hospital potential for each block group. This step computes the denominator $\sum_{k=1}^{n} \left(S_k d_{ik}^{-\beta} \right)$ in the Huff model. For each block group area indexed by i, it has a unique value.

Step 5. Calibrating the Probability of Each Hospital Being Visited by a Block Group: Add a Joiner node (Node 13) and link Math Formula (Node 11) and GrougBy (Node 12) to it. In the dialog of Joiner (Node 10), (1) under Joiner settings tab, choose Origin ID for both Top Input ('left' table) and Bottom Input ('right' table); (2) under Column Selection, choose all columns in the Include box for Top Input (left table), choose Sum(Potent) in the Include box for Bottom Input (right table), and keep other default settings. Click OK and execute it.

Add two Math Formula nodes and link Joiner (Node 13) → Math Formula (Node 14) → Math Formula (Node 15). Configure them to append two new columns Prob and EstFlow by the expressions, $Potent$/$Sum(Potent)$ and $Popu2020$*$Prob$, respectively. This step completes the calibration of probability

$P_{ij} = S_j d_{ij}^{-\beta} / \sum_{k=1}^{n} \left(S_k d_{ij}^{-\beta} \right)$, and multiplying it by the corresponding population in a block group area yields the predicted patient flow from a block group area i to a hospital j.

Step 6. Identifying Hospitals Being Visited with the Highest Probability: Add a GroupBy node (Node 16) and link Math Formula (Node 15) to it. In the dialog, (1) under the Groups tab, choose Origin ID for Group column(s) and Aggregation method (column name) for Column naming; (2) under the manual Aggregation tab, choose Prob for Column and Maximum for Aggregation. Click OK and execute it. The resulting table includes a field called Max*(Prob), which represents the maximum probability for each block group area to visit a specific hospital.

Add a Joiner node (Node 17), and link GroupBy (Node 16) and Math Formula (Node 15) to it. In the dialog of Joiner (Node 17), (1) under Joiner settings tab, choose Origin ID and Max*(Prob) for Top Input ('left' table), and select Origin ID and Prob for Bottom Input ('right' table); (2) under Column Selection, keep empty in the Include box for Top Input (left table), choose all columns in the Include box for Bottom Input (right table), and keep other default settings. Execute it. This node identifies the hospitals being visited by each block group area with the highest probability. It is worth noting that the combination of GroupBy and Joiner nodes can be highly beneficial for identifying the minimum or maximum value for each group.

Step 7. Defining service areas by the Huff model: Add a Joiner node (Node 18) and link GeoFile Reader (Node 2) and Joiner (Node 17) to it. In the dialog of Joiner (Node 17), (1) under Joiner settings tab, choose OBJECTID for Top Input ('left' table) and Origin ID for Bottom Input ('right' table); (2) under Column Selection tab, keep all columns in the Include box for Top Input (left table), choose Destination ID and Prob in the Include box for Bottom Input (right table), and keep other default settings. Execute it.

Add a Geospatial View node (Node 19) and link Joiner (Node 18) to it. In the dialog, choose Destination ID for Marker color column and CartoDB PositronNoLabels for base map. Figure 4.4a shows the hospital service areas derived by the Huff model.

Step 8. Mapping the probability surface for a hospital's visitation: The service area of a hospital in Figure 4.4a encompasses census block groups whose residents visit that hospital with the highest probability. This step uses one hospital OLTLRMC as an example to demonstrate the calibration and mapping of its varying probabilities of being visited by residents across the study area.

Add a Row Filter node (Node 20) and link Math Formula (Node 15) to it. In the dialog, keep Include rows by attribute value checked, choose Destination ID for Column to test, check use pattern matching and input 5 for it, and keep other default settings. Click OK and execute it. This keeps all the records for the hospital OLTLRMC.

Link GeoFile Reader (Node 2) and Row Filter (Node 20) to Joiner (Node 18). In the dialog of Joiner (Node 18), (1) under Joiner settings tab, choose OBJECTID for Top Input ('left' table) and Origin ID for Bottom Input ('right' table); (2) under Column Selection, choose all columns in the Include box for Top Input (left table), choose Prob in the Include box for Bottom Input (right table), and keep other default settings. Execute it.

In the dialog of Geospatial View (Node 19), choose Prob for Marker color column to visualize the probabilities of residents visiting OLTLMC (Figure 4.4b). Users

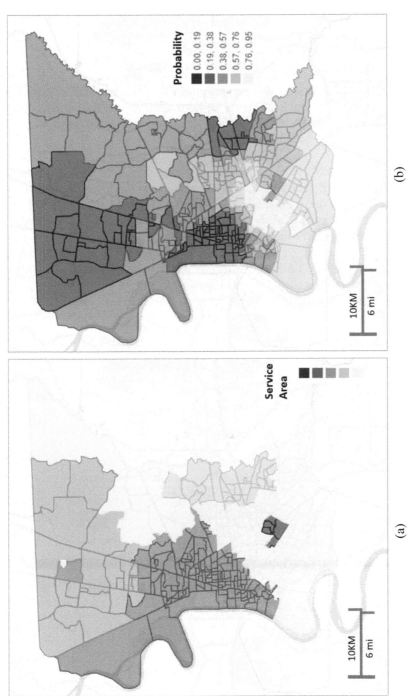

FIGURE 4.4   Huff model results: (a) hospital service areas, and (b) probabilities of choosing OLTLMC

may choose any other hospital and repeat the process to map its probabilities being visited across the study area.

4.2   CASE STUDY 4B: AUTOMATED DELINEATION OF HOSPITAL SERVICE AREAS IN FLORIDA

This case study uses toolkits developed by Python Script node to delineate the hospital service areas in Florida. It uses three methods: Huff model, Dartmouth method, and Network Community Detection methods. The folder Florida contains all data for Case Study 4B listed as follows:

1) subfolder FL_HSA contains a polygon feature class ZIP_Code_Area. zip for 983 ZIP code areas in Florida, and a table for hospitalization volumes between these ZIP code areas OD_All_Flows.csv with 37,180 non-zero flows and a total service volume of 2,392,066 (refer to Case Study 2B on the hospitalization table details),

2) a CSV file FLplgnAdjAppend.csv is a spatial adjacency matrix, and another CSV file FLAdjUpdate.csv is the updated spatial adjacency matrix generated in Step 3 of Subsection 4.2.1, and

3) Python scripts in Network Community Detection Method for Python Script.py and Dartmouth Method for Python Script.py.

Figure 4.5 shows the workflow for Part 1 of this case study.

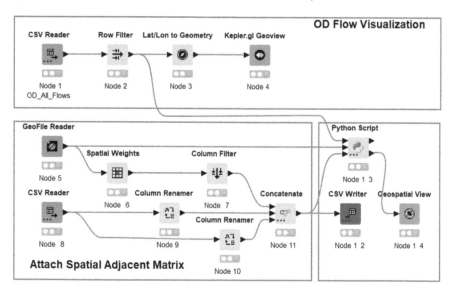

FIGURE 4.5   Workflow of Part 1 of Case Study 4B

4.2.1 Part 1: Delineating HSAs by the Refined Dartmouth Method

Step 1. Visualizing patient hospitalization OD flows with Kepler.Gl Geoview: Create a new workflow Case4B1. Add a CSV Reader node (Node 1) to the Editor and input the file path for OD flow, e.g., E:\KNIME\CMGISV3\Chp04\data\OD_All_Flows.csv. Execute it.

Add a Row Filter node (Node 2) and link CSV Reader (Node 1) to it. In the dialog, check Include rows by attribute value, choose AllFlows for Column to test, check use range checking, input 1 for lower bound, and keep other default settings. Click OK and execute it to filter out the records with AllFlows=0 (n=37180).

Add a Lat/Lon to Geometry node (Node 3) and link Row Filter (Node 2) to it. In the dialog, choose YCoord for Latitude column and XCoord for Longitude column, keep epsg:4326 for CRS, and execute it to append the residential points.

Add a Kepler.gl Geoview node (Node 4) and link Lat/Lon to Geometry (Node 3) to it. Execute it directly. Right click and choose Interactive View: Geospatial View to open the configure window.

In the window (Figure 4.6a), (1) click the hidden menu button ![icon] to expand the menu, (2) click the drop down menu on the state to expand the layer menu, (3) click Polygon under Basic to bring out the visualization type menu, (4) click Arc (or Line), and then choose YCoord for Source Lat, XCoord for Source Lon, Hosp_Y_MC for Target Lat, and Hosp_X_MC for Target Lon to view the OD flow map, (5) click ⋮ on Color to expand the pane, choose AllFlows for Color based on, quantile for Color Scale, (6) click ⋮ on Stroke to expand the pane, and choose AllFlows for Stroke based on and sqrt for Stroke Scale. Figure 4.6b shows the visualization of OD flows between the residential areas and hospitals. Click ![icon] to view the OD flows in 3D, as shown in Figure 4.6c.

Step 2. Building Spatial Adjacency Matrix: Add a GeoFile Reader node (Node 5) and input the file path for ZIP Code Area data, e.g., E:\KNIME\CMGISV3\Chp04\data\ZIP_Code_Area.zip. Execute it.

Under ![icon] Exploratory Spatial Data Analysis, drag ![icon] Spatial Weights to the Editor and link GeoFile Reader (Node 5) to it. In the dialog of Spatial Weights (Node 6) (Figure 4.7a), choose ZoneID for Column ID, Queen for Weights category, and keep other default settings. Click OK and execute it to generate spatial weight table. The resulting table from the Spatial Weights node contains three columns: focal, neighbor, and weight (Figure 4.7b). It is important to note that both the columns focal and neighbor contain values that are identical to ZoneID.

Users can change the spatial weight type under the Weights category, which generally contains two categories: contiguity-based and distance-based. The former includes Rook, Queen, and their order options shown in Figure 4.8a,[1] and the latter can be further divided into two subsets: Distance Band (binary values for Binary

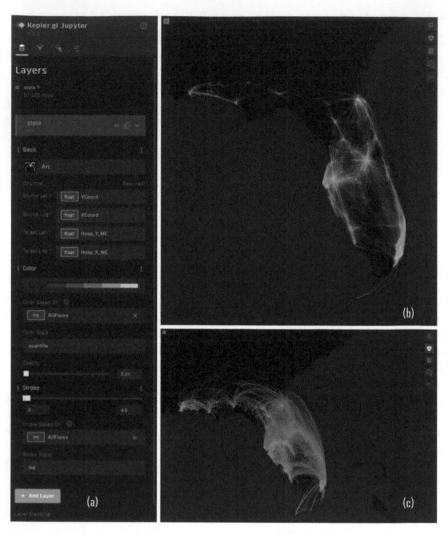

FIGURE 4.6 Visualization of OD flow between residential areas and hospitals: (a) Parameter pane, (b) 2D Map, and (c) 3D Map

Distance Band, K Nearest) and Distance Function (Inverse Distance and Kernel) shown in Figure 4.8b. A Lattice spatial weight can be generated by inputting the number of rows and columns shown in Figure 4.8c.

The relationship between the number of `neighbor` (n=5) and the `weight` value (w=1/n=0.2) can be easily observed from the first five rows of the result table (Figure 4.7b). To filter out the column weight for the subsequent spatial weight list concatenating, add a Column Filter (Node 7) and link it to the Spatial Weights

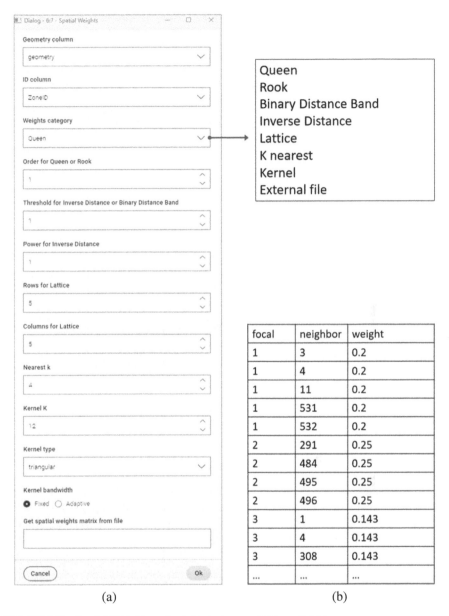

FIGURE 4.7   Spatial Weights node: (a) interface and (b) result table

(Node 6). In the dialog, keep `focal` and `neighbor` in the Include box, click OK, and execute it.

Step 3. Adjusting Spatial Adjacency Matrix: In the previous step, only ZIP code areas that share edges or corners were considered as neighbors, while some areas may be separated by natural barriers such as rivers or lakes but connected by man-made

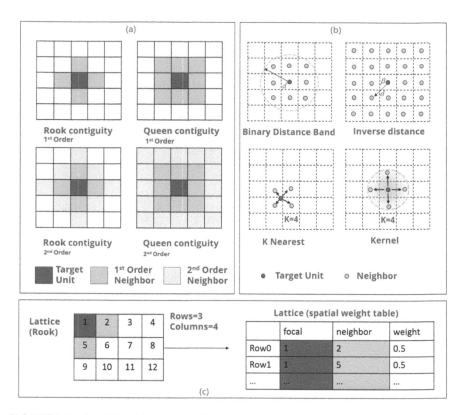

FIGURE 4.8   Spatial weights: (a) Contiguity-based; (b) Distance-based, (c) Lattice

infrastructures like roads or ferries. Thus, the spatial adjacency matrix needs to be adjusted to account for such connectivity.

To do this, add a CSV Reader node (Node 8) to the workflow, input the file path for the additional spatial adjacency matrix to amend connectivity, e.g., E:\KNIME\ CMGISV3\Chp04\data\FLplgnAdjAppend.csv, uncheck Has column name, and execute it. The resulting table contains pairs of dual neighboring units, and its column names (Column0 and Column1) are different from the generated spatial weight table.

To address the discrepancies in column names, add two Column Renamer nodes and link CSV Reader (Node 8) to them. In Column Renamer (Node 9), rename Column0 to focal, Column1 to neighbor; in Column Renamer (Node 10), rename Column0 to neighbor, Column1 to focal. The concatenation of these two tables generates a directional spatial neighboring list.

Under Manipulation > Row > Transform, drag Concatenate to the Editor. Click the function button . . . on the icon of Concatenate (Node 11) and choose Add input port to add the third port to the node (Figure 4.9a). Link Column Filter (Node 7) and Column Renamer (Nodes 9 and 10) to the three input ports of Concatenate (Node 11), and execute it with the default settings to combine the three tables.

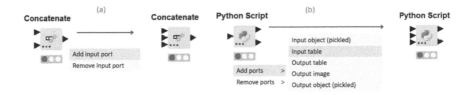

FIGURE 4.9   Adjusting ports of (a) Concatenate node and (b) Python Script node

To export the updated spatial adjacency matrix, add a CSV Writer node (Node 12). Link Concatenate (Node 11) to it, and save it as a CSV file named FLAdjUpdate.csv with a proper local file path, e.g., E:\KNIME\CMGISV3\ Chp04\data\LAdjUpdate.csv.

Step 4. Using the Dartmouth Method to Delineate HSAs: Add a Python Script node to the workflow and click on the function button . . . on the node icon to add an additional input port by selecting Add Ports and then Input Table (Figure 4.9b). Repeat this step to set up a total of 3 input ports and 2 output ports for Python Script (Node 13).

Link Row Filter (Node 2), GeoFile Reader (Node 5) and Concatenate (Node 11) to the three input ports of Python Script (Node 13), which correspond to OD flows, ZIP Code Areas and the adjusted spatial weight table. In the Python Script, copy and paste the code from the Dartmouth Method for Python Script.py file and execute the node. Users can refer to Figure 3.17 in Section 3.3 for a reference on the layout of the R nodes, which shares a similar interface with the Python Script node.

This Python script implements the refined Dartmouth method for defining HSAs. It is important to note that the input polygon feature should have a projected coordinate system to ensure proper calibration of geographic compactness. The default values for threshold size and LI (Localization Index) are 1,000 and 0.5, respectively. If a threshold size is used, the input polygon feature must contain a population field with non-zero values.

The Python Script node produces two outputs. The bottom output generates 136 contiguous HSAs and calculates several regionalization indices within the new HSAs. The top output adds a field HSAID to the input feature to identify which ZIP code areas are contained within each HSA. The bottom resulting data contains fields COUNT_ZoneID and Num_DZoneID, which record the number of ZIP code areas and the number of destination ZIP code areas (with hospitals) within each HSA, respectively. Additionally, four fields, LI, POPU, Compactness and EstTime, represent the localization index, population size, geographic compactness, and average travel time of each HSA.[2] Figure 4.10 displays the compactness of HSAs in Florida delineated by the Dartmouth Method.

Users can customize the Dartmouth Method by adjusting the minimum localization index and population constraints by revising the code threshold-Size = int(1000), miniLocalIndex = 0 in the Python Script node. This generates a new set of HSAs based on the updated criteria.

To visualize the output HSAs, add a Geospatial View node (Node 14) to the workflow and connect the bottom output port of Python Script (Node 13) to it. The

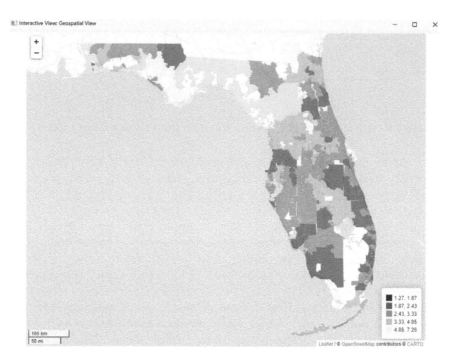

FIGURE 4.10 HSAs with Geographic Compactness values in Florida delineated by Dartmouth Method

Geospatial View allows users to interactively explore the HSAs and their attributes, such as the localization index (Figure 4.18a), population size, and geographic compactness (Figure 4.10).

Step 5. Optional: Integrating the Huff Model and Dartmouth Method: Both the Huff model and Dartmouth method use the plurality rule by assigning a ZIP code area to a hospital being visited most. However, the Dartmouth method is based on actual patient flows, and the Huff model is based on estimated flows in absence of such data. One may use the Huff model to first estimate OD flows and then apply the Python script of Dartmouth method as a post treatment on the estimated flows to derive HSAs automatically.

Users can copy the node cluster from Case 4A (Nodes 11–17 in steps 4–6) to calculate the estimated flows (EstFlow) based on the population column POPU, hospital capacity NUMBEDS, and travel time Total_Time_min. Revise the code edgeFlowField = "AllFlows" in the Python Script to edgeFlow-Field = "EstFlow" and re-execute Node 13. The bottom output port generates a feature class that contains the dissolved HSAs with predicted total patient volumes. As the Huff model is prone to derive spatially non-contiguous HSAs, one may impose additional constraints such as threshold size to merge small "islands" to their adjacent larger HSAs.

4.2.2 PART 2: DELINEATING HSAS BY SPATIALIZED NETWORK COMMUNITY DETECTION METHODS

Figure 4.11 shows the workflow for this part and its sub-workflow in the component Parameters.

Step 6. Enabling column selection for flow variables: Create a new workflow Case4B2. Similar to Step 1, add CSV Reader (Node 1) and GeoFile Reader (Node 2) to the Editor to read the files OD_All_Flows.csv and ZIP_Code_Area.zip, respectively.

In the following, we leverage the function of component and the nodes in Configuration of Workflow Abstraction to build an interface for Spatialized Network Community Detection Methods.

Under Workflow Abstraction > Configuration > Selection, drag two Column Selection Configuration nodes to the Editor, and link GeoFile Reader (Node 2) to the two nodes. In the dialog of Column Selection Configuration (Node 3) shown in Figure 4.12, enter Input OriginID for Label and OriginID for Parameter/ Variable Name, choose Dropdown for Selection Type, check Number(long),

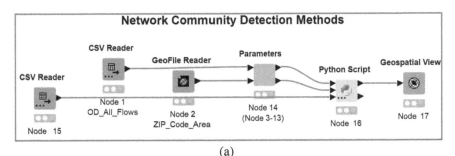

(a)

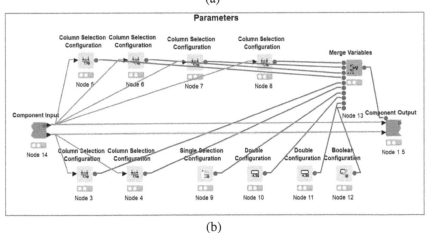

(b)

FIGURE 4.11 Workflow of Part 2 of Case Study 4B: (a) overall workflow, and (b) sub-workflow in the component Parameters

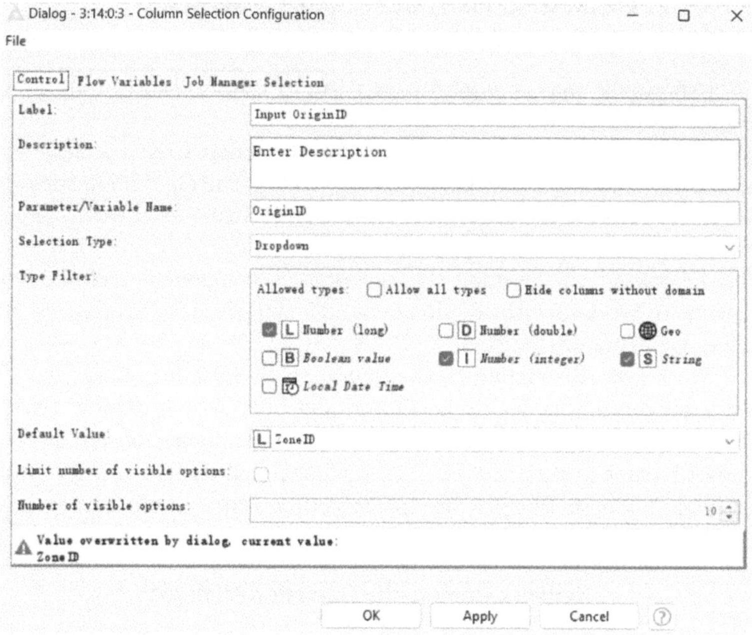

FIGURE 4.12 Interface of Column Selection Configuration

`Number(integer)` and `String` for Type Filter, and choose `ZoneID` for Default value. Click OK and execute it. This node builds a dropdown menu with label and explanation so users can define the flow variable `OrgininID` from the input data `ZIP_Code_Area`.

To set the dialog of Column Selection Configuration (Node 4), follow the pattern of Node 3 dialog, and input the parameters listed in Table 4.1 for Node 4. Additionally, add four more Column Selection Configuration nodes (Nodes 5–8) and link them to the CSV Reader (Node 1). Configure these nodes to input the parameters as specified in Table 4.1.

Step 7. Enabling parameter adjustment for value selections: Under Workflow Abstraction> Configuration> Selection, drag Single Selection Configuration (Node 9) to the Editor. In the dialog shown in Figure 4.13, input `Network Community Detection Model` for Label and `Model` for Parameter/Variable Name, `Dropdown` for Selection Type, input `ScLeiden` and `Sclouvain` in two lines for Possible Choices, and choose `ScLeiden` as the Default Value. Click OK and execute it. This enables the user to choose one of the two strings `ScLeiden` or `Sclouvain` as the value of flow variable Model.

Under Workflow Abstraction > Configuration > Input, drag Double Configuration (Node 10) to the Editor. In the dialog (refer to Figure 3.3), input `Resolution` for Label, `Resolution` for Parameter/Variable Name, and 20 for Default Value. Click OK and execute it. This node allows the user to define a flow variable `Resolution` with a default value of 20.

TABLE 4.1

Parameters in the Dialog of Column Selection Configuration Nodes

Node#	Label	Parameter/Variable Name	Type Filter*	Default Value
3	Input OriginID	OriginID	ID	ZoneID
4	Input Population	OriginPopulation	Value	POPU
5	ODmatrix OriginID	ODmatrixOriginID	ID	PatientZipZoneID
6	ODmatrix DestinationID	ODmatrixDestinationID	ID	Hosp_ZoneID
7	ODmatrix Flow	ODmatrixFlow	Value	AllFlows
8	ODmatrix Cost	ODmatrixCost	Value	Total_Time_min

Source: * ID type: Number (integer), Number (Long), Number (double); Value type: Number (integer), Number (Long), String

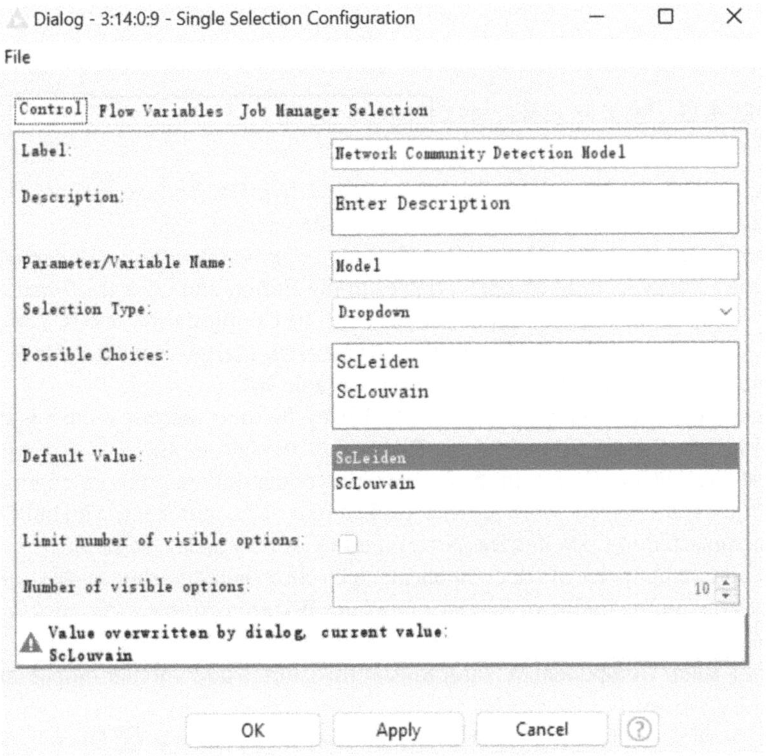

FIGURE 4.13　Interface of Single Selection Configuration

Add another Double Configuration node (Node 11). In the dialog, input `Population constraint` for Label, `constraint` for Parameter/Variable Name, and `1000` for Default Value. Click OK and execute it. This node allows the user to define a flow variable `constraint` with a default value `1000`.

Add a Boolean Configuration node (Node 12). In the dialog (Figure 4.14), input `Apply Spatial Constraint` for Label and `ApplySC` for Parameter/Variable

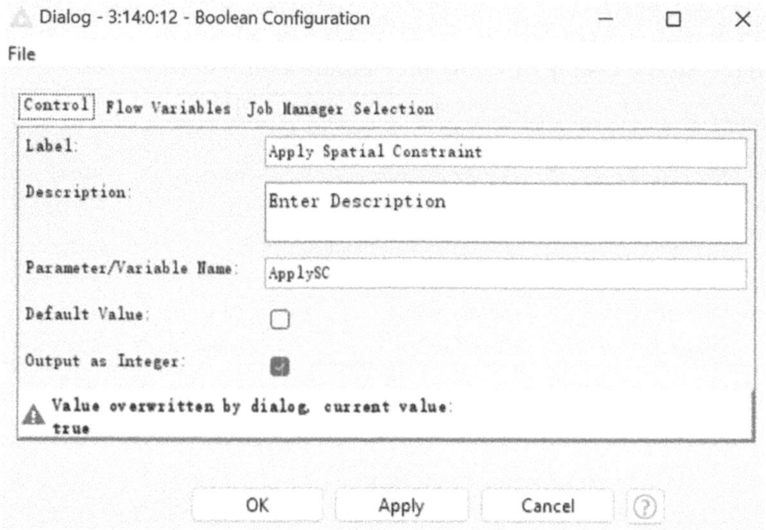

FIGURE 4.14 Interface of Boolean Configuration

Name, and check `Output as Integer`. Click OK and execute it. This node defines a True/False variable `ApplySc` with the value of `1/0`.

Step 8. Merging variables and configuring component settings: Under Workflow Control > Variable, drag Merge Variables to the Editor, and click the functional button . . . on the node to add input ports for the 10 Configuration nodes. Link Nodes 3–12 to Merge Variables and execute it. Right click Merge Variables (Node 13) and choose Merged variables to view the flow variable list.

Next, select all nodes (Node 3-Node 13) by holding the left mouse button and drawing a rectangle around them. Right-click on one of the selected nodes and choose Create Component. In the Create Component dialog, input `Parameters` as the name. The selected nodes are then packed into one component, with built-in input ports connected to CSV Reader (Node1) and GeoFile Reader (Node 2).

To add output ports of the component, right-click on it and choose ▦ Component > ▦ Setup. In the dialog of Setup Component Wizard (Figure 4.15), click `Add`; and in the dialog of Add Meta Port, choose Table to add `Out_1(Table)` to the Out Ports of Setup Component Wizard. Repeat this step to add another output table port `Out_2(Table)`.

Right click on the component and choose ▦ Component > ▦ Open to enter the Editor canvars of component Parameters. There are two new nodes inside: Component Input (Node 14) and Component Output (Node 15). Component Input (Node 14) is already connected to the Configuration nodes, with its two ports representing the output data of CSV Reader (Node 1) and GeoFile Reader (Node 2). Connect the two output ports of Component Input to the two input ports of Component Output, which transfer the data downstream to other workflows and nodes.

FIGURE 4.15   Interface of Setup Component Wizard

Link Merge Variables (Node 13) to the input Flow Variable Port of Component Output (Node 15) to pass the variables as well. Double-click the Component Output node to open the dialog and choose all variables in the Include box (Figure 4.16). Click OK and execute it. Figure 4.11b shows the nodes configuration in the component `Parameters`.

The Component Input and Component Output dialogs are useful for importing variables from upstream workflows or exporting variables to downstream workflows. They act as a valve for global variables in the entire workflow and local variables within the component.

Go back to the workflow `Case4B2` in the Editor and double-click the component `Parameters` to bring up the user-defined interface pane for parameter adjustment, as shown in Figure 4.17.

Step 9. Delineating HSAs by ScLeiden method: Add a CSV Reader node (Node 15) and read the file of updated spatial adjacency matrix, e.g., `E:\KNIME\ CMGISV3\Chp04\data\FLAdjUpdate.csv`.

Add a Python Script node (Node 16), link the two output ports of the component Parameters to the top two input ports of Python Script (Node 16), and link CSV Reader (Node 15) to the third input port. In the dialog of Python Script (Node 16), input the Python code from `Network Community Detection Method for Python Script.py` and execute the node. The bottom output port generates contiguous Health Service Areas (HSAs) and calculates several regionalization indices in the newly derived HSAs, and the top output port generates a layer by adding a field to the input feature and identifying which ZIP code areas are contained in what HSA.

The code uses flow variables for its parameters, such as `inputIDField = knio.flow_variables['Origin ID']`. Users can modify the parameters in

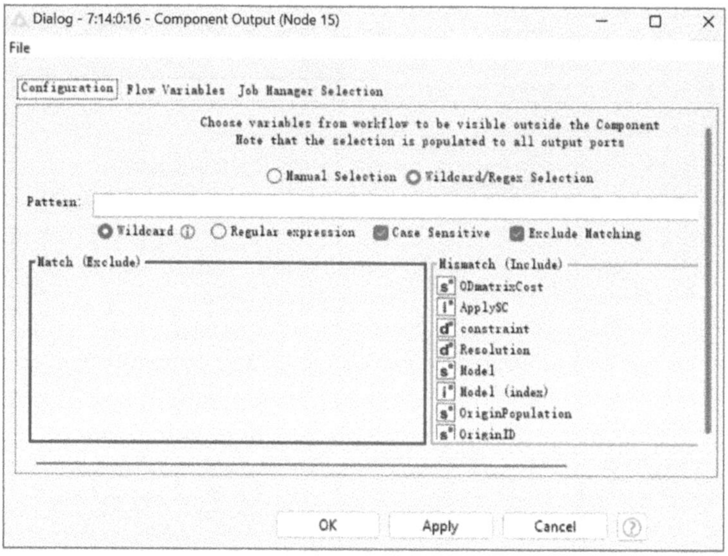

FIGURE 4.16 Interface of Component Output

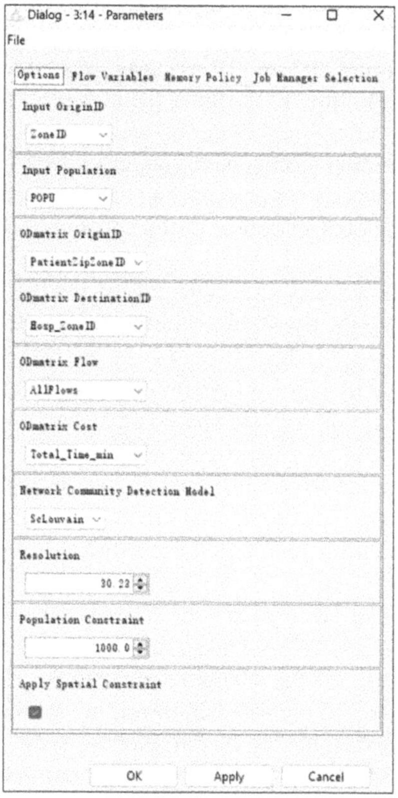

FIGURE 4.17 Interface pane created by component Parameters

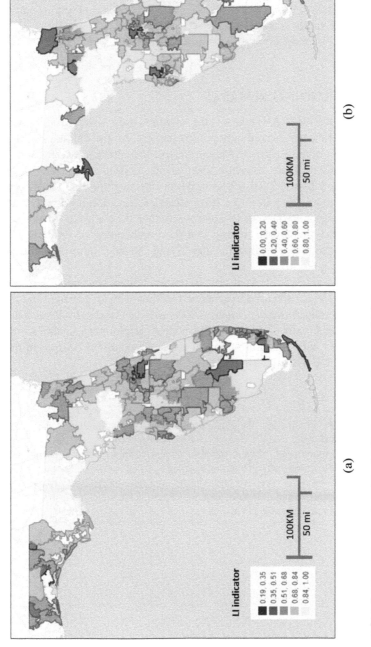

FIGURE 4.18   LI values in HSAs: (a) Dartmouth method, and (b) ScLouvain method

the panel without changing the code in Python Script. This implements a customized automation tool for Spatialized Network Community Detection Methods.

The output contains 138 contiguous HSAs, each with fields HSAID for each HSA id, COUNT_ZoneID and Num_DZoneID for numbers of (origin) ZIP code areas and destination ZIP code areas (or hospitals) within each HSA, respectively, and four additional fields LI, POPU, Compactness and EstTime. Figure 4.18b shows the variation of LI values across HSAs.

Experiment with different resolution values or switch the Delineation Method to see the impacts.

4.3 CONCLUDING REMARKS

For built-in KNIME AP nodes, this chapter uses several nodes in IO and Manipulation that are covered in previous chapters. Additional new nodes are introduced. For example, *Scripting (Python Script)* is applied to use Python packages. Other nodes like *Column Selection Configuration*, *Single Selection Configuration*, and *Merge Variables* are utilized to build an interface for components, which can be seen as customized nodes. The flow variables are configured in the Component Input and Component Output to control whether they are passed downstream to other workflows. It is worth noting that the combination of GroupBy and Joiner nodes can be highly beneficial for identifying the minimum or maximum ID or value for each group.

In terms of Geospatial Analytics Extension, three new nodes are introduced. *Kepler.gl Geoview* is first used to generate OD flow with arcs or lines in both 2D and 3D maps. Secondly, *Voronoi (Thiessen Polygon)* node is used to define supply-based proximal areas. Finally, *Spatial Weights* node in Exploratory Spatial Data Analysis (ESDA) is used here to capture polygon adjacency and will be used in most of the nodes in ESDA and Spatial Modeling in Chapter 8.

NOTES

1 Contiguity-Based Spatial Weights, https://geodacenter.github.io/workbook/4a_contig_weights/lab4a.html
2 Localization index (LI) refers to the ratio of service flows within a HSA divided by the total service flows originated from the same HSA. POPU is the total population in a region. Compactness is the corrected perimeter-to-area ratio or *PAC* $= P / (3.54 * A^{0.5})$, where *P* denotes perimeter and *A* denotes area. EstTime is the weighted average travel time for patients originated from the HSA.

5 GIS-Based Measures of Spatial Accessibility and Application in Examining Healthcare Disparity

The chapter introduces a family of GIS-based methods for defining spatial accessibility. Accessibility refers to the relative ease by which the locations of activities, such as work, school, shopping, recreation, and healthcare, can be reached from a given location. Disadvantaged population groups (e.g., low-income and minority residents) often suffer from poorer accessibility, and the disparity calls for planning and public policy to mitigate the problem. The popular *Two-Step Floating Catchment Area (2SFCA) method* and the *Generalized 2SFCA (G2SFCA) method* measure spatial accessibility for residents to reach a certain service or activity. The *inverted 2SFCA (i2SFCA) method* estimates potential crowdedness in facilities. The Two-Step Virtual Catchment Area (2SVCA) method measures virtual accessibility (accessibility via internet), e.g., telehealth.

Case Study 5 illustrates several accessibility measures for primary care physicians in Baton Rouge, Louisiana. Part 1 describes step-by-step implementations of conventional 2SFCA and gravity-based 2SFCA methods. Part 2 details the 2SVCA method for measuring telehealth accessibility. Part 3 uses a looping workflow to implement sensitivity analysis with various parameter settings.

5.1 PART 1: MEASURING ACCESSIBILITY OF PRIMARY CARE PHYSICIANS IN BATON ROUGE

This case study extends the work reported in Wang, Vingiello and Xierali (2020), and uses the same study area: nine parishes in the Baton Rouge Metropolitan Statistical Area (MSA). Updates include using more recent population data and adding the implementation of a virtual accessibility measure. The case study uses Euclidean distances to measure travel impedance d_{ij} for simplicity so that we can focus on implementing the accessibility measures in GIS.

The population data is extracted from the 2020 census as illustrated in Case Study 1 in section 1.3 of Chapter 1. Data of physicians (including specialties and geographic locations) in Louisiana in 2022 are extracted from the Doctors and Clinicians National Downloadable File released by the Centers for Medicare and Medicaid Services (CMS). There are 594 primary care physicians (PCP) aggregated to 172 locations (census blocks). The internet infrastructure data is based on the Fixed Broadband Deployment Block Data (dated December 31, 2020) via the FCC website (*www.fcc.gov/general/broadband-deployment-data-fcc-form-477*).

DOI: 10.1201/9781003304357-5

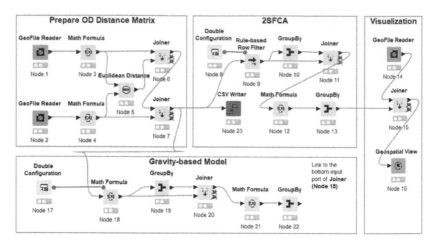

FIGURE 5.1 Workflow of Part 1 of Case Study 5

The FCC broadband data contains maximum upload and download speed records by different service providers (consumer vs. business).

The following features in the data folder BRMSA are used for this project:

1. polygon features MSAbkg.zip (n = 574) and MSAcounty.zip (n = 9) are 2020 census block groups and counties, respectively, for reference and mapping,
2. polygon feature MSAbkg1 and point feature MSAbkgPt.zip (n = 570 for both) are 2020 census block groups and corresponding centroids after excluding four census block groups with no residents, and contain fields GEOID for unique geography identifier ID, POP100 for population, and MaxAdDown, MaxAdUp and bandR for mean fixed broadband download speed, mean upload speed and for subscription rate of broadband of consumer service providers, respectively,
3. point feature PCPmsa.zip contains fields PCP for number of FTE (full-time equivalent) primary care physicians (PCP), and MaxAdDown and MaxAdUp for mean fixed broadband download and upload speeds by business service providers, in 2020 census block centroids, and
4. OD Cost Matrix table ODwithBroadband.csv is generated by Step 1, and contains travel cost between MSAbkgPt and PCPmsa based on the Euclidean distance represented by the field Distance, with fields Origin ID and Destination ID representing population and physician location ids, respectively. It also contains broadband information.

Figure 5.1 shows the workflow for Part 1 of this case study.

5.1.1 IMPLEMENTING THE 2SFCA METHOD

Recall the definition for 2SFCA, for each demand location i, search all supply locations (j) that are within the threshold distance (d_0) from location i (i.e., catchment area i), and sum up the supply to demand ratios R_j at those locations to obtain the accessibility A_i at demand location i:

$$A_i = \sum_{j \in \{d_{ij} \leq d_0\}} R_j = \sum_{j \in \{d_{ij} \leq d_0\}} \left(\frac{S_j}{\sum_{k \in \{d_{kj} \leq d_0\}} D_k} \right), \tag{5.1}$$

where d_{ij} is the distance between i and j, and R_j is the supply to demand ratio at supply location j that falls within the catchment centered at i (i.e., $d_{ij} \leq d_0$). A larger value of A_i indicates a better accessibility at a location.

Step 1. Computing distances between population and physician locations: Create a new workflow Case5A. Add two GeoFile Reader nodes to the Editor to read the file of population locations (MSAbkgPt.zip) and physician locations (PCPmsa. zip), e.g., E:\KNIME\CMGISV3\Chp05\data\MSAbkgPt.zip for Node 1 and E:\KNIME\CMGISV3\Chp05\data\PCPmsa.zip for Node 2.

Add two Math Formula nodes, link GeoFile Reader (Node 1) to Math Formula (Node 3), and link GeoFile Reader (Node 2) to Math Formula (Node 4). In both Math Formula nodes, use the expression $$ROWINDEX$$+1 to append an Integer column DID for demand ID in Math Formula Node 3, and an Integer column SID for supply ID in Math Formula Node 4. Be sure to check Append Column and Convert to Int in the dialogs.

Add a Euclidean Distance node (Node 5) and link the two Math Formula nodes to it. In the dialog, input DID for Origin ID column, SID for Destination ID column, choose Kilometer for Distance unit. Execute it to calculate the distances between the two sets of locations. The resulting table contains columns Origin ID, Destination ID and Distance indicating the IDs of the two input datasets and the distances between them in kilometers. The distance table has $570 \times 172 = 98,040$ records.

Step 2. Attaching population and physician data to the distance table: Add two Joiners nodes (Nodes 6 and 7), and link Math Formula (Node 3) and Euclidean Distance (Node 5) to the Joiner (Node 6). In the dialog of Joiner (Node 6), (1) under Joiner settings tab, choose DID for Top Input ('left' table) and Origin ID for Bottom Input ('right' table), check making integer types compatible; (2) under Column Selection, keep GEOID, MaxAdDown, MaxAdUp, POP100 and bandR in the Include box of both Top Input (left table) and all columns in the Include box of Bottom Input (right table); and (3) keep other default settings. Execute it. This step attaches the key attributes of population locations to the OD (origin-destination) distance list.

Link Joiner (Node 6) and Math Formula (Node 4) to Joiner (Node 7). In the dialog of Joiner (Node 7), (1) under Joiner settings tab, choose Destination ID for Top Input ('left' table) and SID for Bottom Input ('right' table), and check making integer types compatible; (2) under Column Selection, keep all columns in the Include box of both Top Input (left table) and choose PCP, MaxAdDown, MaxAdUp in the Include box of Bottom Input (right table); and (3) keep other default settings. Execute it. This step attaches the attributes of physician locations to the OD distance list. In the result table, the two columns, MaxAdDown (right) and MaxAdUp (right) are for the duplicated column names of PCPmsa.

Step 3. Extracting distances within a catchment area: Similar to Step 5 in Section 3.1 of Chapter 3, we use a flow variable to enable parameter control. Add a Double Configuration node (Node 8) and Rule-based Row Filter (Node 9) to the workflow editor, link Joiner (Node 7) to Rule-based Row Filter (Node 9), and link

Double Configuration (Node 8) to the left upper corner of Rule-based Row Filter (Node 9) as a flow variable.

In the dialog of Double Configuration (Node 8), input `Threshold for Catchment Area` for Label, `Threshold` for Parameter/Variable Name, and `32.18` for Default value (32.180 kilometers for 20 miles). Click OK and execute it. This step attaches the flow variable `Threshold` with a value of 32.18.

In the dialog of Rule-based Row Filter (Node 9), input the expression as `$Distance$ <= $${DThreshold}$$=> TRUE`. Click OK and execute the node. The output table only includes those distances within 20 miles, and thus implements the selection conditions $i \in \{d_{ij} \leq d_0\}$ and $k \in \{d_{kj} \leq d_0\}$ in Equation 5.1.

Step 4. Summing up population around each physician location: Add a GroupBy node (Node 10) and link Rule-based Row Filter (Node 9) to it. In the dialog of GroupBy (Node 10), (1) under the Groups tab, choose `Destination ID` for Group column(s) and `Aggregation method (column name)` for Column naming, (2) under Manual Aggregation tab, choose `POP100` for Column and `Sum` for Aggregation. Click OK and execute it. The resulting column `Sum(POP100)` is the total population within the threshold distance from each physician location, and thus calibrates the term $\sum_{k \in \{d_{kj} \leq d_0\}} D_k$ in Equation 5.1.

Step 5. Computing initial physician-to-population ratio at each physician location: Add a Joiner node (Node 11), and link GroupBy (Node 10) and Rule-based Row Filter (Node 9) to it. In the dialog of Joiner (Node 11), (1) under Joiner settings tab, choose `Destination ID` for both Top Input ('left' table) and Bottom Input ('right' table); (2) under Column Selection, choose `Sum(POP100)` in the Include box of Top Input (left table) and choose all columns in the Include box of Bottom Input (right table); and (3) keep other default settings. Execute it.

Add a Math Formula node (Node 12) and link Joiner (Node 11) to it. Configure it to append a new column `docpopR` with the expression `1000*$PCP$/$Sum(POP100)$`. This assigns an initial physician-to-population ratio to each physician location. This step computes the term $\dfrac{S_j}{\sum_{k \in \{d_{kj} \leq d_0\}} D_k}$ in Equation 5.1. The ratio is inflated 1,000 times to indicate the physician availability per 1,000 residents.

Step 6. Summing up physician-to-population ratios by population locations: Add a GroupBy node (Node 13) and link Math Formula (Node 12) to it. In the dialog of GroupBy (Node 13), (1) under the Groups tab, choose `Origin ID` for Group column(s) and `Keep original name(s)` for Column naming, (2) under Manual Aggregation tab, choose `docpopR` and `GEOID` for Column, `Sum` and `First` for Aggregation, respectively. Click OK and execute it. The column `docpopR` represents the total availability of physicians reachable from each residential location, and thus completes the calibration of accessibility in Equation 5.1.

Step 7. Mapping accessibility: To visualize the results, add a GeoFile Reader node (Node 14) to read the census block group units from the file `MSAbkg.zip`, e.g., `E:\KNIME\CMGISV3\Chp05\data\MSAbkg.zip`.

Add a Joiner node (Node 15) and link GeoFile Reader (Node 14) and GroupBy (Node 13) to it. In the dialog of Joiner (Node 15), (1) under Joiner settings tab, choose `GEOID` for both Top Input ('left' table) and Bottom Input ('right' table), (2) under Column Selection, choose all columns in the Include box of both Top Input (left table) and Bottom Input (right table), and keep other default settings. Execute the node to join the `docpopR` to the census block group polygon feature based on their common fields `GEOID`.

Add a Geospatial View node (Node 16) and link Joiner (Node 15) to it. In the dialog, choose `docpopR` for Marker color column, and `FisherJenks` for Classification Method to visualize the accessibility scores.

Step 8. Optional: Sensitivity analysis using various threshold distances: A sensitivity analysis can be conducted to examine the impact of using different threshold distances. For instance, the study can be repeated through steps 3–7 by using threshold distances of 15, 10 and 5 miles, and results can be compared. More on sensitivity analysis will be discussed in Part 3.

5.1.2 IMPLEMENTING THE GRAVITY-BASED 2SFCA METHOD

Recall the definition of Gravity-Based 2SFCA Method, written as:

$$A_i = \sum\nolimits_{j=1}^{n} \frac{S_j d_{ij}^{-\beta}}{V_j}, \text{where } V_j = \sum\nolimits_{k=1}^{m} D_k d_{kj}^{-\beta}, \tag{5.2}$$

where n and m are the total numbers of supply and demand locations, respectively, and the other variables are the same as in Equations 5.1. Compared to Equations 5.1, where only S and D within a catchment range enter into the formulation, Equation 5.2 discounts S and D by a distance decay effect in the form of a power function (i.e., gravity model).

The process of implementing the gravity-based 2SFCA method is similar to that of the conventional 2SFCA method. The differences are highlighted below.

Step 9. Computing population potential for each physician location: Similar to Step 3, add a Double Configuration node (Node 17) and a Math Formula node (Node 18) to the Editor, link Joiner (Node 7) to Math Formula (Node 18), and link Double Configuration (Node 17) to the left upper corner of Math Formula (Node 18) as a flow variable.

In the dialog of Double Configuration (Node 17), input `Distance friction coefficient` for Label, `Beta` for Parameter/Variable Name, and `1.0` for Default value (assuming $\beta = 1.0$). Configure Math Formula (Node 18) to append a new column `PPotent` with the expression `$POP100$* ($Distance$^(-1*$$ {DBeta}$$))`.

Similar to Step 4, add a GroupBy node (Node 19) and link Math Formula (Node 18) to it. In the dialog of GroupBy (Node 19), (1) under the Groups tab, choose `Destination ID` for Group column(s) and `Aggregation method (column name)` for Column naming, (2) under Manual Aggregation tab, choose `PPotent` for Column and `Sum` for Aggregation. Click OK and execute it. The column

Sum(PPotent) calibrates the term $V_j = \sum_{k=1}^{m} D_k d_{kj}^{-\beta}$ in Equation 5.2, which is the population potential for each physician location.

Step 10. Computing accessibility attributable to individual physician locations: Similar to step 5, add a Joiner node (Node 20) and link GroupBy (Node 19) and Math Formula (Node 18) to it. In the dialog of Joiner (Node 20), (1) under Joiner settings tab, choose Destination ID for both Top Input ('left' table) and Bottom Input ('right' table), (2) under Column Selection, choose Sum(PPotent) in the Include box of Top Input (left table) and choose all columns in the Include box of Bottom Input (right table), and (3) keep other default settings. Execute it.

Add a Math Formula node (Node 21) and link Joiner (Node 20) to it. Configure it to append a new column docpopR with the expression 1000*PCP*($Distance$^

(-1*$${DBeta}$$))/$Sum(PPotent)$. This computes the term $\dfrac{S_j d_{ij}^{-\beta}}{V_j}$ in

Equation 5.2. Again, a multiplier 1000 is applied to avoid small values.

Step 11. Summing up accessibility to all physician locations: Similar to step 6, copy a GroupBy node (Node 13) and paste it as a new node GroupBy (Node 22) with its pre-configured settings. Link Math Formula (Node 21) to it and execute it directly (refer to step 6 for detailed settings). The column docpopR is the gravity-based accessibility in Equation 5.2.

Link GroupBy (Node 22) to the bottom input port of Joiner (Node 15) and re-execute the workflow. Right-click Geospatial View (Node 16) and choose Interactive View: Geospatial View to see the visualization of gravity-based accessibility. Similarly, the sensitivity analysis for the gravity-based method can be conducted by changing the β value in Double Configuration (Node 17).

To facilitate the subsequent analysis on virtual accessibility in Part 2, add a CSV Writer node (Node 23) and link Joiner (Node 7) to it. In the dialog, specify the file path (e.g., E:\KNIME\CMGISV3\Chp05\data\ODwithBroadband.csv). Execute the node to write the data to the local disk.

5.2 PART 2: IMPLEMENTING THE 2SVCA METHOD

The *generalized 2SFCA (G2SFCA) method* is formulated as:

$$A_i = \sum_{j=1}^{n} \left[S_j f(d_{ij}) / \left(\sum_{k=1}^{m} D_k f(d_{kj}) \right) \right] \tag{5.3}$$

where a general distance decay function $f(d_{ij})$ or $f(d_{kj})$, replacing the power function $d_{ij}^{-\beta}$ or $d_{kj}^{-\beta}$ in Equation 5.2, can take various forms.

The 2SVCA for virtual accessibility is formulated as

$$VA_i = a_i \sum_{j \in (d_{ij} \leq d_0)}^{n} \left[a_j S_j f(b_j) / \sum_{k \in (d_{kj} \leq d_0)}^{m} \left(a_k D_k f(b_k) \right) \right] \tag{5.4}$$

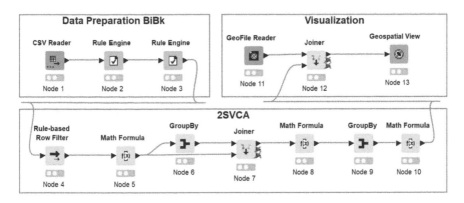

FIGURE 5.2 Workflow of Part 2 of Case Study 5

where the degree of a facility S_j and a demand location D_k participating in telehealth services between them is a function of their broadband strengths $f(b_j)$ and $f(b_k)$, in place of the distance decay function $f(d_{kj})$ or $f(d_{ij})$ in Equation 5.3.

There are three additional parameters associated with broadband subscription rates: (1) a_j is applied on facility S_j to reflect the portion with commercial broadband subscription ($a_j = 1$ is assumed for all primary care physician (PCP) facilities in this case study), (2) a_k is applied on demand D_k to capture that only those residents with consumer broadband subscription would contribute to telehealth services offered by S_j, and (3) a_i is applied to discount the initial virtual accessibility score assigned to demand location i because only this portion of residents have the consumer broadband subscription. In this case study, if the minimum download and download speeds in a block group are ≥ 25 and 3 mbps, respectively, it is considered with broadband access (i.e., $f(b_i)$ or $f(b_k) = 1$); and otherwise, no access (i.e., $f(b_i)$ or $f(b_k) = 0$).

Figure 5.2 shows the workflow for Part 2 on implementing the 2SVCA method.

Step 1. Defining broadband availability for population and physician locations: Create a new workflow `Case5B`, and add a CSV Reader node (Node 1) in the Editor. In the dialog of CSV Reader (Node 1), input the file path for `ODwithBroadband.csv` exported in the last step of Part 1. Click OK to confirm the settings and execute the node to read the data. The resulting data include the columns: `GEOID, MaxAdDown, MaxAdUp, POP100, bandR, Origin ID, Destination ID, Distance, PCP, MaxAdDown (right)` and `MaxAdUp (right)`.

Next, we will use the broadband speed requirements to define broadband availability.

Add a Rule Engine node (Node 2) to the Editor and link CSV Reader (Node 1) to it. In the dialog of Rule Engine (Node 2) shown in Figure 5.3, input two lines in the

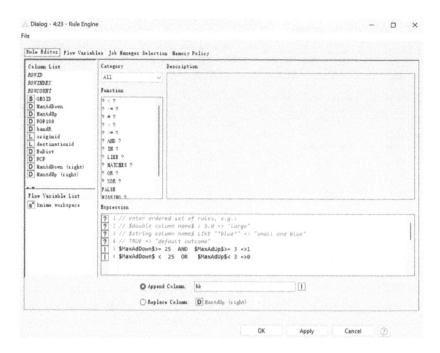

FIGURE 5.3 Interface of Rule Engine

Expression by double-clicking the column names MaxAdDown and MaxAdUp in Column List and the appropriate operation symbols in Function:

```
$MaxAdDown$>= 25 AND $MaxAdUp$>= 3 =>1
$MaxAdDown$ < 25 OR $MaxAdUp$< 3 =>0
```

Check Append Column and input bk as the column name. Click OK and execute it. This node assigns the value 1 or 0 to the column bk according to the values of MaxAdDown and MaxAdUp of demand (population) locations.

Add another Rule Engine node (Node 3) and link Rule Engine (Node 2) to it. Append a new column bi with the following expression:

```
$MaxAdDown (right)$>= 25 AND $MaxAdUp (right)$>= 3 =>1
$MaxAdDown (right)$ < 25 OR $MaxAdUp (right)$< 3 =>0
```

This node assigns the value 1 or 0 to the column bi according to the values of MaxAdDown(right) and MaxAdUp(right) of supply (physician) locations.

Step 2. Attaching broadband availability data to the OD matrix: Similar to step 3 of Section 5.1, add a Rule-based Row Filter node (Node 4) and link Rule Engine (Node 3) to it. In the dialog of Rule-based Row Filter (Node 4), revise the expression to $Distance$ <= 32.18=> TRUE and check Include TRUE matches.

Click OK and execute the node. The distance table includes only those distance pairs within 20 miles and thus implements the selection conditions $j \in \{d_{ij} \leq d_0\}$ and $k \in \{d_{kj} \leq d_0\}$ in Equation 5.4.

Step 3. Summing up population with broadband subscription in areas with available broadband services around each physician location: Similar to Step 9 of Section 5.1, add a Math Formula node (Node 5) and link Rule-based Row Filter (Node 4) to it. Configure it to append a new column wpop with the expression $bandR$*$POP100$*bk, which is the value $a_k D_k f(b_k)$ in Equation 5.4. Only locations with broadband services are eligible for participating in telehealth, and the population there is further discounted by (limited to) those with broadband subscriptions.

Add a GroupBy node (Node 6) and link Math Formula (Node 5) to it. In the dialog of GroupBy (Node 6), (1) under the Groups tab, choose Destination ID for Group column(s) and Aggregation method (column name) for Column naming, (2) under Manual Aggregation tab, choose wpop for Column and Sum for Aggregation. Click OK and execute it. The field Sum(wpop) is the total population having adequate broadband access within the threshold distance from each physician location, and therefore implements calibration of the term $\sum_{k \in (d_{kj} \leq d_0)}^{m} (a_k D_k f(b_k))$ in Equation 5.4.

Step 4. Computing initial physician-to-population ratio at each physician location: Similar to step 10 of Section 5.1, add a Joiner node (Node 7) and link GroupBy (Node 6) and Math Formula (Node 5) to it. In the dialog of Joiner (Node 7), (1) under Joiner settings tab, choose Destination ID for both Top Input ('left' table) and Bottom Input ('right' table), (2) under Column Selection, choose Sum(wpop) in the Include box of Top Input (left table) and all columns in the Include box of Bottom Input (right table), and (3) keep other default settings. Execute it.

Add Math Formula (Node 8) and link Joiner (Node 7) to it. Configure it to append a new column docpopR with the expression 1000*PCP*bi/$Sum(wpop)$. This assigns an initial physician-to-population ratio to each physician location, where only physicians in locations with broadband services and subscribed to broadband contribute to telehealth access. This step computes the term $\dfrac{a_j S_j f(b_j)}{\sum_{k \in (d_{kj} \leq d_0)}^{m} (a_k D_k f(b_k))}$

in Equation 5.4. Again, the ratio is inflated 1,000 times to indicate the physician availability per 1,000 residents.

Step 5. Summing up physician-to-population ratios by population locations: Similar to step 11 of Section 5.1, add a GroupBy node (Node 9) and link Math Formula (Node 8) to it. In the dialog of GroupBy (Node 9), (1) under the Groups tab, choose Origin ID for Group column(s) and Keep original name(s) for Column naming, (2) under Manual Aggregation tab, choose docpopR, bandR and GEOID for Column, and Sum, First and First for Aggregation, respectively. Click OK and execute it.

The column docpopR in the output port represents the virtual accessibility for physicians from each residential location. It includes five locations with zero values due to inadequate access to the internet. This computes the term

$\sum_{j\in(d_{ij}\le d_0)}^{n} \dfrac{a_j S_j f(b_j)}{\sum_{k\in(d_{kj}\le d_0)}^{m}(a_k D_k f(b_k))}$ in Equation 5.4, which is a preliminary virtual

accessibility value that is limited to those with broadband subscriptions there.

Add a Math Formula node (Node 10) and link GroupBy (Node 9) to it. Configure the node to replace the value of column docpopR with the expression $docpopR$*$bandR$ and checking Replace Column. Click OK and execute it.

By multiplying the broadband subscription rates, the Math Formula node returns the overall virtual accessibility score VA_i associated with residential area i in Equation 5.4.

Step 6. Mapping accessibility: Add three nodes, GeoFile Reader (Node 11), Joiner (Node 12) and Geospatial View (Node 13), and configure them with the same settings in step 7 in Section 5.1 to read the census block group units from the file MSAbkg.zip. Link Math Formula (Node 10) to the bottom input port of Joiner (Node 12) to join docpopR to the census block group polygon feature based on their common fields GEOID and re-execute the workflow. Right-click Geospatial View (Node 13) and choose Interactive View: Geospatial View to see the visualization of virtual accessibility.

5.3 PART 3: SENSITIVE ANALYSIS FOR MEASURING ACCESSIBILITY BY WORKFLOW AUTOMATION

This part automates the workflow for 2SFCA with adjustable parameters to implement sensitivity analysis. Figure 5.4 shows the workflow and its sub-workflow in the component 2SFCA.

5.3.1 STANDARD COMPONENT FOR 2SFCA METHOD

Step 1. Standardizing column names of population and physicians: Create a new workflow Case5C. Copy the CSV Reader (Node 1) in Case5B to the Editor to read the data ODwithBroadband.csv.

To standardize the column names for the other two key variables (population and physicians), add five Column Selection Configuration nodes (Nodes 2–6) for users to specify the five key columns, Demand ID, Demand (population), Supply ID, Supply (physicians) and Distance.

Link CSV Reader (Node 1) to Column Selection Configuration (Node 2). In the dialog, (1) input Choose Demand Population for Label and Demand for Parameter/Variable Name, (2) for Type Filter, uncheck Allow all types, check Number(double), Number(integer) and Number (Long), (3) choose POP100 for Default Value, and keep other default settings. Click OK and execute it.

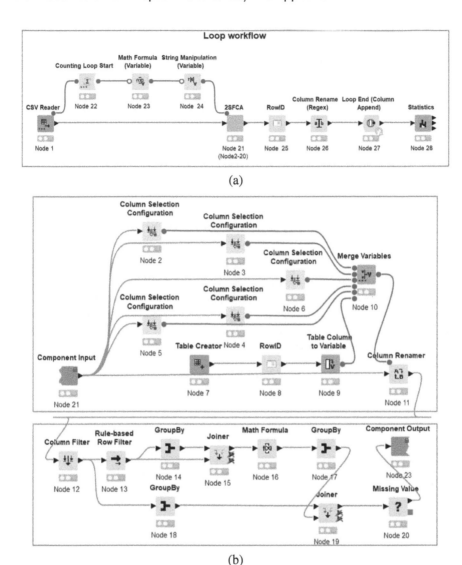

FIGURE 5.4 Workflow of Part 3 of Case Study 5: (a) Loop workflow, (b) workflow in component 2SFCA

This allows the user to choose the name of one numeric column for the Flow Variable Demand with a default value POP100.

Similarly, link CSV Reader (Node 1) to the other 4 Column Selection Configuration nodes, and configure the other 4 nodes according to the settings in Table 5.1.

Step 2. Creating multiple flow variables by Table Creator: The Table Creator node is used to create five flow variables for the string names. Under IO > Other >

TABLE 5.1

Parameters in the Dialog of Column Selection Configuration Nodes

Node#	Label	Parameter/ Variable Name	Type Filter	Default Value
2	Choose Demand Population	Demand	Number(double) Number(integer)	POP100
3	Choose Supply Capacity	Supply	Number (Long)	PCP
4	Choose Distance	Distance		Distance
5	Choose Demand ID	DemandID	Number(integer)	Origin ID
6	Choose Supply ID	SupplyID	String	Destination ID

Modular Data Generation, drag Table Creator to the Editor. In the dialog of Table Creator (Node 7) shown in Figure 5.5a, input `Demandname`, `Supplyname`, `DemandIDname`, `SupplyIDname` and `Distancename` in the first column, `Demand`, `Supply`, `DemandID`, `SupplyID` and `Distance` in the second column, and keep other settings as default. Click OK and execute it. We can also revise the column name and type by double clicking the column header to bring out the dialog of New settings for column as shown in Figure 5.5b.

Under Manipulation > Row > Other, drag RowID to the Editor and link Table Creator (Node 7) to it. In the dialog of RowID (Node 8) shown in Figure 5.6, choose `column1` for New RowID column, check `Remove selected column`, and keep other default settings. Click OK and execute it. This node replaces the row IDs with the value of `column1`.

The RowID node can be used to reset or append row IDs. For example, choosing <none> for New RowID column will reset the ID, while checking `Create new column with the RowID values` and inputting a name for "New column name" will extract the row IDs as a new column.

Under Workflow Control > Variable, drag Table Column to Variable to the Editor, and link RowID (Node 8) to it. Execute Table Column to Variable (Node 9) directly. It builds a flow variable list with the values of row ID, first column for the names and values of the variables, e.g., variable `Supplyname` with a string value of `Supply`.

Add a Merge Variable node (Node 10) and configure it to have 6 input ports, link all 5 Column Selection Configuration nodes (Nodes 2–6) and Table Column to Variable (Node 9) to the six ports. Execute it to merge variables.

Step 3. Renaming and filtering columns with flow variables: Now all the variables can be used for standardizing columns. Add a Column Renamer node (Node 11) and link CSV Reader (Node 1) to its input data port, link Merge Variable (Node 10) to its flow variable port. In the dialog of Column Renamer shown in Figure 5.7a, click Add column to add 5 input box pairs, choose any of the columns

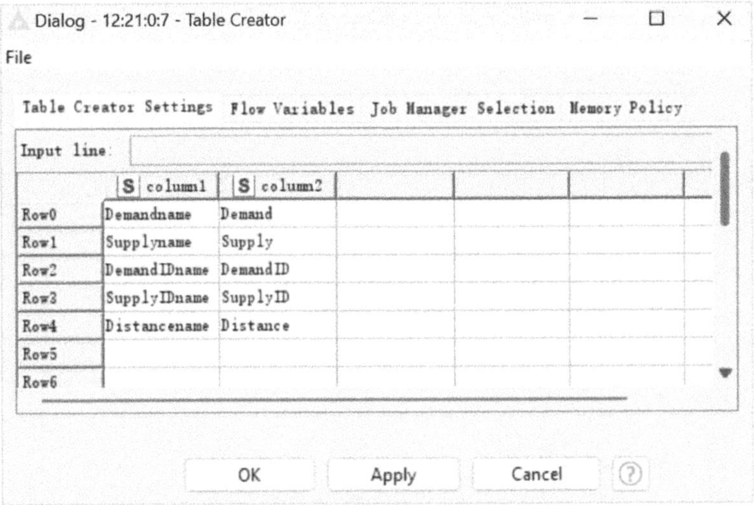

(a)

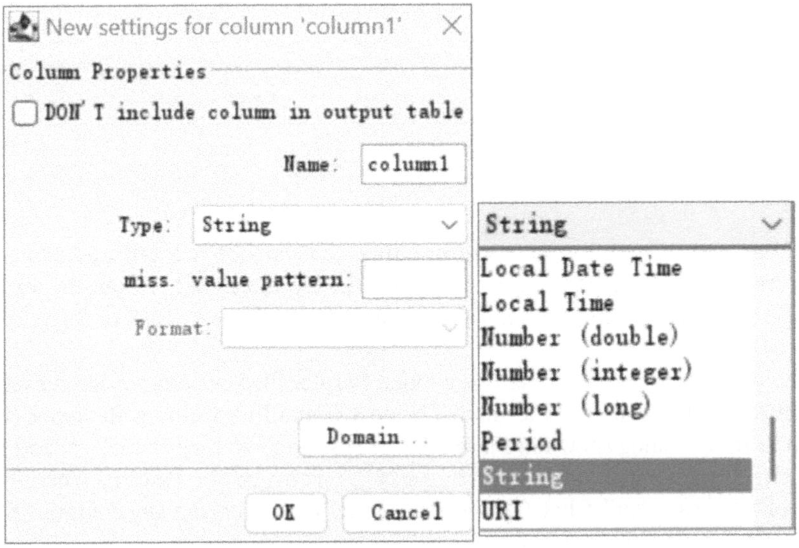

(b)

FIGURE 5.5   Interface of (a) Table Creator and (b) its dialog of column settings

in Column and input any strings in New Name without duplicated names, click OK to accept it. Then right click the node and choose Configure Flow Variables to bring out the Flow Variables dialog. As shown in Figure 5.7b, expand all columns, set `Demand`, `Supply`, `DemandID`, `SupplyID` and `Distance` for old-Name, and `Demandname`, `Supplyname`, `DemandIDname`, `SupplyIDname`

FIGURE 5.6 Interface of RowID

and `Distancename` for the corresponding newName. Click OK and execute it. It automatically displays the variable values in the boxes rather than the variable names. This renames the columns according to Table 5.1, e.g., `PCP` to `Supply` and `POP100` to `Demand`.

The regular expression (Regex) is applied to filter the key columns for subsequent workflow. Add a Column Filter node (Node 12) and link Column Renamer (Node 11) to it. In the dialog shown in Figure 5.8, check `Regex expression`, and input `DemandID|SupplyID|Demand|Supply|Distance` for Pattern, wherein, the symbol "|" means "or". Click OK and execute it. This filters out any columns that do not match the specified pattern.

Step 4. Reconfiguring 2SFCA node cluster: Copy Nodes 9–13 for 2SFCA in the workflow `Case5A`, and paste them into the workflow `Case5C`. They are automatically reordered as Nodes 13–17. Link Column Filter (Node 12) to Rule-based Row Filter (Node 13) to connect the workflows. As some variable names changed, we need to reconfigure Nodes 13–17 (refer to Nodes 9–13 in the workflow `Case5A`).

In the dialog of Rule-based Row Filter (Node 13), revise the Expression as `$Distance$ <= 32.18=> TRUE`. We run with this value first and will come back to revise this expression after finishing setting the threshold value as flow variable.

In the dialog of GroupBy (Node 14), (1) under the Groups tab, keep `SupplyID` for Group column(s), `Aggregation method (column name)` for Column

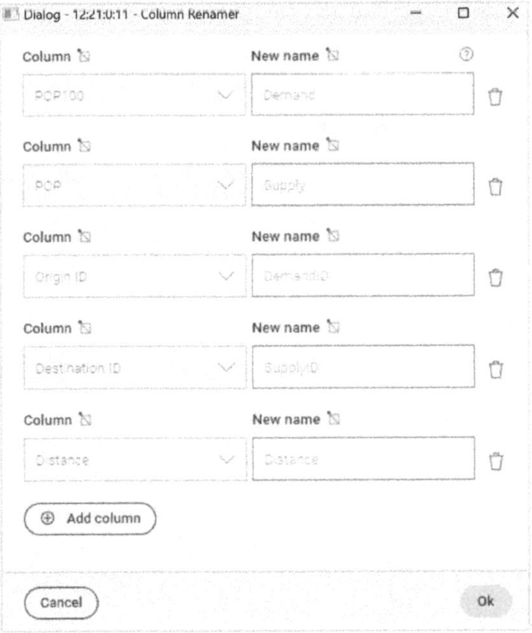

(a)

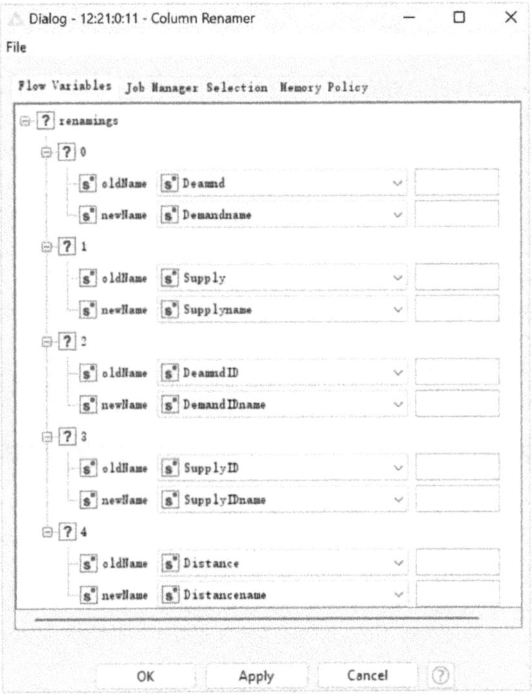

(b)

FIGURE 5.7 Interfaces of (a) Column Renamer, and (b) its Flow Variables

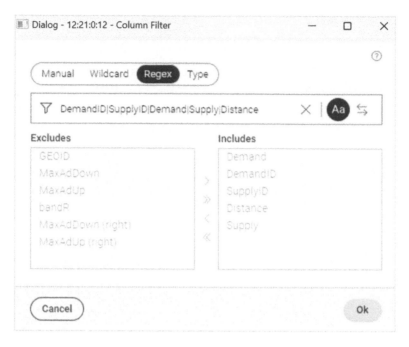

FIGURE 5.8 Interface of Column Filter based on Regex

naming, (2) under Manual Aggregation tab, choose `Demand` for Column and `Sum` for Aggregation. Click OK and execute it.

In the dialog of Joiner (Node 15), (1) under Joiner settings tab, choose `SupplyID` for both Top Input ('left' table) and Bottom Input ('right' table), (2) under Column Selection, choose `Sum(Demand)` in the Include box of Top Input (left table) and choose all columns in the Include box of Bottom Input (right table), and (3) keep other default settings. Click OK and execute it.

In the dialog of Math Formula (Node 16), change the expression to `1000*$Supply$/$Sum(Demand)$`.

In the dialog of GroupBy (Node 17), (1) under the Groups tab, keep `DemandID` for Group column(s), `Keep original name(s)` for Column naming, (2) under Manual Aggregation tab, choose `docpopR` for Column and `Sum` for Aggregation.

Execute the whole workflow to run 2SFCA with the threshold value 32.18 km.

Step 5. Joining results to input population data: As the threshold value increases, the 2SFCA model may exclude certain OD pairs in the OD list because the demand points are beyond the threshold distance of physicians. This can result in a change in the length of the 2SFCA result table, which can cause the looping workflow to fail. To address this issue, we need to assign a zero value to these excluded points.

First, we will rebuild a full list of demand points. Add a GroupBy node (Node 18) and link Column Filter (Node 12) to it. In the dialog, input `DemandID` in the Group Column, Keep original name(s) for Column naming, leave blank for Manual Aggregation. Execute it to generate a table with a single column DemandID (n=570).

Add a Joiner node (Node 19), and link GroupBy (Node 18) and GroupBy (Node 17) to it. In the dialog of Joiner (Node 19), (1) under Joiner settings tab, choose DemandID for both Top Input ('left' table) and Bottom Input ('right' table), check Matching rows and Left unmatched rows to set the node for left join mode; (2) under Column Selection, keep DemandID in the in the Include box of Input (left table), check Wildcard/Regex Selection and Regular expression for the Bottom Inputs ('right' table), and input docpopR for Pattern. Click OK and execute it to join the two tables and only keep the two columns, DemandID and docpopR.

To address missing values for accessibility score docpopR (e.g., no supply within the catchment area of a demand location). Add Missing Value (Node 20) and link Joiner (Node 19) to it. In the dialog, choose Fix Value for Number (Double) and input 0 for Value. Click OK and execute it.

Step 6. Creating component 2SFCA: Right-click on Nodes 2–20 and choose Create Component. Name the component as 2SFCA. For configuring the component, refer to Step 8 in Section 4.2.2. Once the component is created, double-click on it to show the interface for the 2SFCA model (Figure 5.9a). Remember to configure the component to have one input data port and one output port.

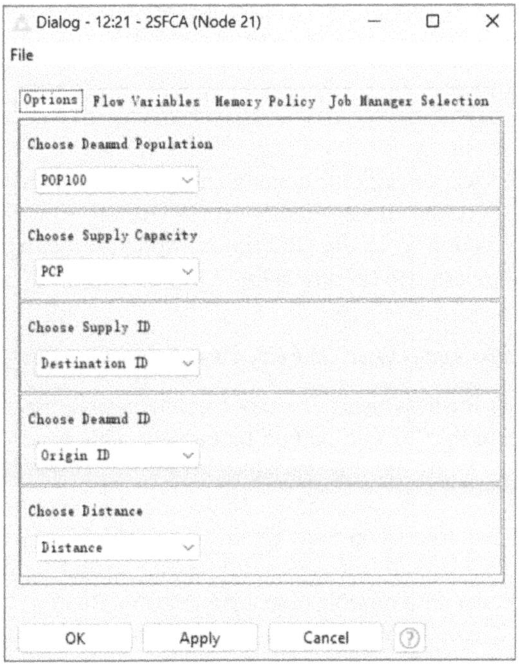

(a)

FIGURE 5.9 Component 2SFCA: (a) Interface and (b) Description editing

(Continued)

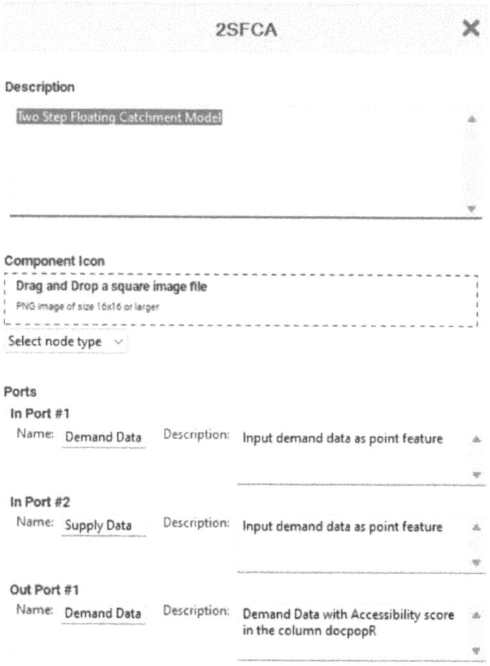

(b)

FIGURE 5.9 Continued

In KNIME AP, users can edit the component as a node for sharing. Right-click on the component 2SFCA and choose Component > Open to view the inside workflow, then click the edit button ✏ in the Description to revise the explanation or add an icon to make it more intuitive (Figure 5.9b).

5.3.2 Sensitive Analysis with 2SFCA Component and Workflow Loop

Sensitivity analysis involves using various catchment area sizes (or threshold distances) to implement the 2SFCA method repeatedly. This is accomplished with the support of workflow loops and flow variable control.

Step 7. Adding Counting Loop Start: Under ⓥ Workflow Control > ↻ Loop Support, drag ⓘ Counting Loop Start (Node 22) to the Editor, click the functional button . . . and choose Add Pass through port to add the flow variable port to the node. Then link the flow variable output port of CSV Reader (Node 1) to Counting Loop Start (Node 22). In the dialog, input 5 for Number of loops. Right click the node and choose Unaltered Input to view the flow variable table. It has two variables: maxIterations (value: 5) and currentIteration that varies with each time of loop, starting from 0.

Step 8. Using Math Formula (Variable) to build varying threshold value: Under ⚡ Workflow Control > **V** Variables > drag 🔣 Math Formula (Variable) to the Editor, and link Counting Loop Start (Node 22) to it. In the dialog, input 10+($${IcurrentIteration}$$*5) in the Expression, check Append Variable and input thresholdstep for it, and check Convert to Int. Click OK and execute it. Math Formula (Variable) (Node 23) creates the variable thresholdstep with varying integer values from 10 to 30 Kilometers.

Step 9. Using String Manipulation (Variable) to build varying column name: From the previous step, the workflow generates a table with columns DemandID and docpopR, and we need to rename the resulting column docpopR accordingly.

Add a String Manipulation (Variable) (Node 24) node and link Math Formula (Variable) (Node 23) to it. In the dialog, check Append Variable and input AccLoop for it. Input the expression join(string($${Ithreshold-step}$$),"KM"). Click OK and execute it to create the flow variable Accloop to replace the column name docpopR, with the value of 10KM for first iteration of the workflow.

Step 10. Configuring 2SFCA component: Link String Manipulation (Variable) (Node 24) to the input flow variable port of the component 2SFCA. Such a change resets the workflow. Re-execute the component.

Open the component 2SFCA. In the dialog of Component Input, choose all variables in the Include box manually or by checking Wildcard/Regex Selection and Exclude Matching, and keeping empty for Pattern. In the dialog of Rule-based Row Filter (Node 13), revise the Expression as $Distance$ <= $${Dthresholdstep}$$=> TRUE. This step enables the workflow to use the changing value of the flow variable thresholdstep, which is defined in the Math Formula (Variable) node (Node 23).

Step 11. Handling 2SFCA result table for looping: The KNIME workflow can use two types of result combination: Loop End node for vertically concatenating results, and Loop End (Column Append) node for horizontally appending results. The latter is appropriate for the 2SFCA looping. To avoid having duplicate ID columns, we will set DemandID as the row ID.

Add a RowID node (Node 25) and the component 2SFCA to it. In the dialog, check Replace RowID with selected column values or create a new one, choose DemandID for New RowID column, and check Remove selected column. Click OK and execute it. The result table contains only one column docpopR with DemandID as Row ID.

Since all loops generate a column named docpopR, we need to rename and differentiate them with the predefined flow variable AccLoop. Add a Column Rename (Regex) (Node 26) and link RowID (Node 25) to it. In the dialog, keep the default inputs under Main Configuration tab (Figure 5.10a); under Flow Variables tab, choose AccLoop for replaceString (Figure 5.10b). This node renames all (here, only one) columns with the string value of AccLoop (10KM for the column in the first iteration).

Under Workflow Control > Loop Support, drag Loop End (Column Append) to the Editor, and link Column Rename (Regex) (Node 27) to it. Execute the node with

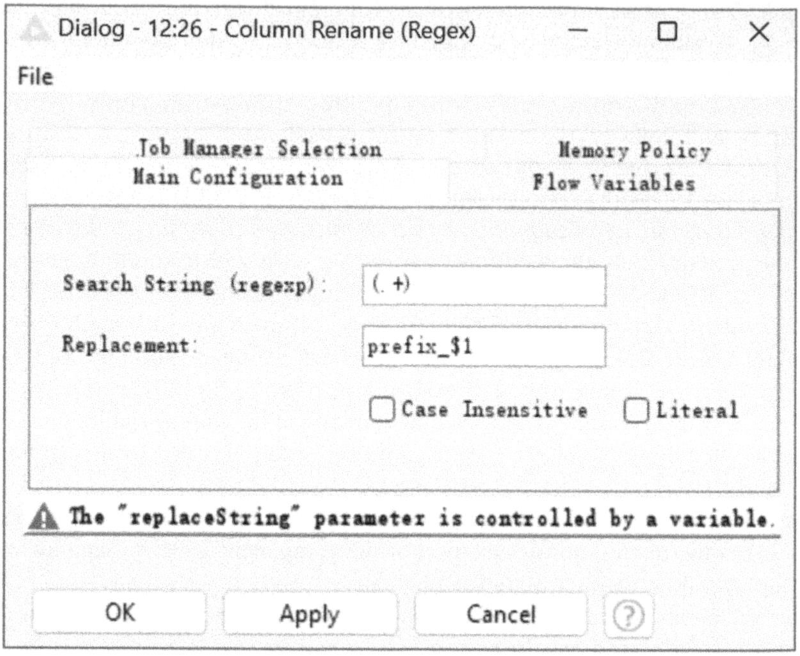

Dialog - 12:26 - Column Rename (Regex)   — ☐ ✕

File

Job Manager Selection   Memory Policy
Main Configuration   Flow Variables

Search String (regexp):   (. +)

Replacement: prefix_$1

☐ Case Insensitive   ☐ Literal

⚠ The "replaceString" parameter is controlled by a variable.

OK   Apply   Cancel   ⑦

(a)

Dialog - 12:26 - Column Rename (Regex)   — ☐ ✕

File

Job Manager Selection   Memory Policy
Main Configuration   Flow Variables

s° searchString

s° replaceString   s° AccLoop

b° isCaseInsensitive

b° isLiteral

OK   Apply   Cancel   ⑦

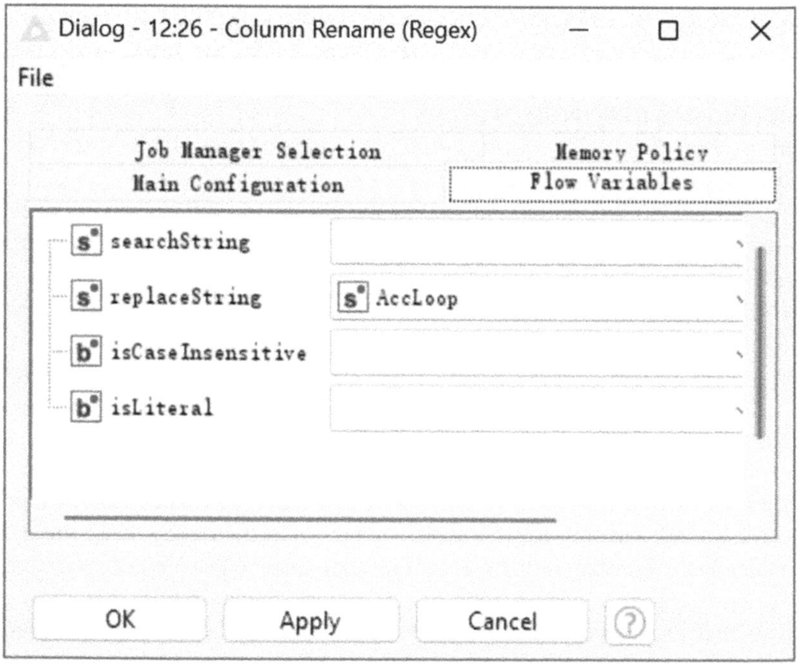

(b)

FIGURE 5.10   Interface of Column Rename (Regex): (a) Main Configuration, and (b) Flow Variables

the default settings to run the loop workflow. The result table records the accessibility scores with different threshold values, ranging from 10KM to 30KM.

To obtain a statistical description for the accessibility scores, add Statistics (Node 28) and link Loop End (Column Append) to it. Execute the node directly with the default settings. The first output port contains the statistical description of the five columns, including Min, Max, Std deviation, etc.

Users can build a loop workflow with a component for the gravity-based 2SFCA model using the same procedures described in Steps 1–9. Users can also implement the inverted 2SFCA (i2SFCA) model by swapping the data sources between the top and bottom ports of the 2SFCA component. See Table 5.1 in the main book for results.

5.4   CONCLUDING REMARKS

This chapter introduces several KNIME built-in nodes, such as *Table Creator* for inputting data like Excel, and *RowID* for creating a new ID or extracting an ID as a new column. The use of flow variables is emphasized in building standardized components, which can be created or modified using nodes like *Double Configuration, Column Selection Configuration,* and *Math Formula (Variable)*. Additionally, *Column Renamer* and *Column Rename (Regex)* nodes can be used for column naming standardization.

For Geospatial Analytics extension nodes, the chapter covers *GeoFile Reader, Geospatial View,* and *Euclidean Distance*. The nodes in Spatial Network can be utilized for calculating OD cost matrix across road networks.

Overall, the chapter highlights the importance of creating standardized components and building loop workflows to facilitate GIS automation.

6 Function Fittings by Regressions and Application in Analyzing Urban Density Patterns

This chapter discusses how to identify the best fitting function to capture urban and regional population density patterns. Such an approach emphasizes the influence of a center or multiple centers on areawide density patterns in a city or across a region. By examining the change of density function over time, one can detect the growth pattern for urban and regional structures. The methodological focus is on function fittings by regressions and related statistical issues.

Chicago has been an important study site for urban studies attributable to the legacy of so-called "Chicago school". The study area is the core six counties (Cook, DuPage, Kane, Lake, McHenry and Will) in Chicago CMSA based on the 2000 census data. The project analyzes the density patterns at both the census tract and survey township levels to examine the possible *modifiable areal unit problem (MAUP)*.

The following features and Python files in the subfolder `ChiUrArea` under the folder `Chicago` are provided:

1. Census tract feature `trt2k.zip` for the larger 10-county MSA region is used to extract census tracts in this 6-county study area (field "`popu`" is the population data in 2000).
2. Feature `polycent15.zip` contains 15 centers identified as employment concentrations from a previous study (Wang, 2000), which includes the Central Business District (CBD) with field $CENT15\_ = 15$.
3. Feature `twnshp.zip` contains 115 survey townships in the study area, providing an alternative areal unit that is relatively uniform in area size.
4. Feature `cnty6.zip` defines the 6-county study area.
5. Three Python script snippet files, `NonlinearRegression.py`, `WeightedOLS.py` and `NonlinearRegressionAssumption3.py`, implement various regression models.

This study implements spatial analysis tasks such as distance computation, areal interpolation, and various regressions such as Ordinary Least Squares (OLS), nonlinear and weighted regressions.

DOI: 10.1201/9781003304357-6

6.1   PART 1: FUNCTION FITTINGS FOR MONOCENTRIC MODELS AT THE CENSUS TRACT LEVEL

Figure 6.1 shows the workflow for implementing the monocentric models.

Step 1. Data preparation: extracting study area and CBD location: Create a new workflow `Case6A`. Begin by adding a GeoFile Reader node (Node 1) and inputting the file path for the Census tract data, e.g., `E:\KNIME\CMGISV3\Chp06\data\trt2k.zip`. Run the node to read the data.

Add a Math Formula node (Node 2) and link the GeoFile Reader (Node 1) to it. Configure it to append a new column `popden` that measures population density in persons per km² with the expression `$POPU$*1000000/$AREA$`.

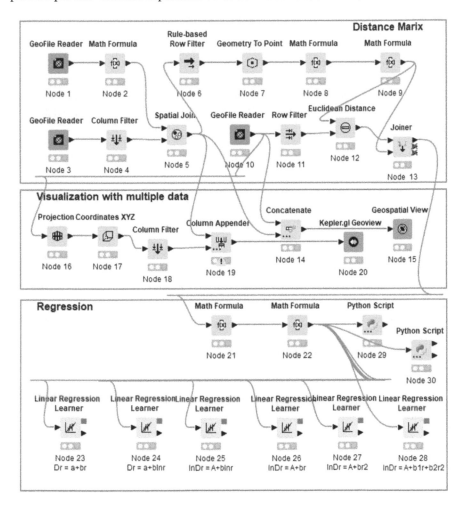

FIGURE 6.1   Workflow of Part 1 of Case Study 6

Add another GeoFile Reader node (Node 3) to read the boundaries of the six counties in the study area, e.g., E:\KNIME\CMGISV3\Chp06\data\cnty6.zip. Add a Column Filter node (Node 4) and keep only geometry in the Include box; or check Type Selection and then Geo (polygon). Click OK and execute it to only keep the geometry column.

Add a Spatial Join node (Node 5), link Math Formula (Node 2) and Column Filter (Node 4) to it. In the dialog, choose Has its center in for Match mode, click OK and execute it to extract 1,837 census tracts in the six-county area.

We need to limit the analysis of population density pattern to tracts with non-zero population. Add a Rule-based Row Filter node (Node 6) and link Spatial Join (Node 5) to it. In the dialog, input $popden$>0=> TRUE for the Expression and keep Include TRUE matches checked. Execute it to eliminates five tracts with zero population.

Add a Geometry to Point node (Node 7) and link Row Filter (Node 6) to it. Execute it directly to transform the data to centroid points. Add two Math Formula nodes (Node 8 and Node 9) and configure Node 8 to append a column lnpopden with the expression ln($popden$). Configure Node 9 to append an integer column NID with the expression $$ROWINDEX$$+1, which will be used for OD list joining.

Step 2. Computing distances between tract centroids and CBD: Add GeoFile Reader (Node 10), and input the file path for Census tract, e.g., E:\KNIME\ CMGISV3\Chp06\data\polycent15.zip. Run it to read the data of centers.

Add Row Filter (Node 11) and link GeoFile Reader (Node 10) to it. In the dialog, check Include row by attribute value, choose CENT15_ for Column to test, and input 5 for use pattern matching. Click OK and execute it to extract the data for CBD.

Add a Euclidean Distance node (Node 12), link Math Formula (Node 9) and Row Filter (Node 11) to it. In the dialog, input NID for Origin ID column, OBJECTID for Destination ID column and Kilometer for Distance unit. Execute it to compute Euclidean distances between tract centroids and the CBD. Note that in the output table, the column Origin ID is for census tract centroids and the column Destination ID is for CBD.

Add a Joiner (Node 13), link Euclidean Distance (Node 12) and Math Formula (Node 9) to it. In the dialog of Joiner (Node 13), (1) under Joiner Settings tab, choose Origin ID for Top Input ('left' table) and NID for Bottom Input ('right' table), (2) under Column Selection, choose Distance in the Include box for Top Input (left table) and keep all columns for Bottom Input (right table), and (3) keep other default settings and execute it. This step attaches the distance values to the feature of census centroids.

Before we continue with the regression task, this section will introduce two ways to visualize multiple data using the Geospatial View node and Kepler.gl View node.

Step 3. Optional: Mapping density with concatenated data and Geospatial View: Add Concatenate (Node 14), link Spatial Join (Node 5) and GeoFile Reader

(Node 10) to it. In the dialog, keep `Union` checked, click OK and execute it directly. The output table contains all columns of the two inputs, with many empty values in most columns except for `geometry`. This can be utilized to visualize multiple geospatial data at one time with the Geospatial View node. For visualizing more data, click the function icon . . . on the Concatenate node and choose `Add input port` to combine more input data. Refer to Step 3 in Section 2.1.3 for more details.

Add Geospatial View (Node 15) and link Concatenate (Node 14) to it. In the dialog, choose `popden` for Marker color column to map the population density surface in the study area. This will display the job centers with the same symbology system, and make it difficult to discern individual centers.

Step 4. Mapping density with Column Appender and Kepler.gl View: Kepler.gl View is used here to better visualize multiple data at the same time. However, as Kepler.gl View only allows one column for geometry type, we use census data as the main geometry data and transform job center data as numerical columns for latitude and longitude.

Add Projection (Node 16) and Coordinates XYZ (Node 17). Link GeoFile Reader (Node 10) $\rightarrow$ Projection (Node 16) $\rightarrow$ Coordinates XYZ (Node 17), and execute them with their default settings, e.g., `epsg:4326` for New CRS, and Replace for Output column in Projection. Two additional columns `x` and `y` are appended for longitude and latitude.

Add Column Filter (Node 18) and link Coordinates XYZ (Node 17) to it. Configure it to keep only the three columns in the output port, `x`, `y` and `CENT15_`.

Add Column Appender (Node 19), link Spatial Join (Node 5) and Column Filter (Node 18) to it. In the dialog, check `Use RowIDs from the selected input table` and input 1 for it. The output table uses Census tract data from Spatial Join (Node 5) as the main table (n = 1837) and appends job center data (n = 15). There are many empty values in the appended columns, `x`, `y` and `CENT15_`.

Add Kepler.gl Geoview (Node 20) and link Column Appender (Node 19) to it. Execute it. Right-click and choose "Interactive View: Geospatial View" to open the visualization window. In the View window, it automatically shows the map of census tracts in the default layer state in the parameter pane.

By clicking the green button +Add `Layer`, add a new layer with a default name `new layer` for visualizing the centers, wherein, choosing `Point` for Basic, `y` for Lat, `x` for Lon, and `CENT15_` for labels.

Click the tab of `new layer` and hold the left mouse button (then ✛ displays) to move `new layer` to the above position of the layer `state`. Adjust other visualization parameters to get a favorable symbology style. The layer can be renamed by double clicking. e.g., `Center`.

Figure 6.2 shows the configuration for points data (on the left) and a map of the population density across census tracts and job centers. Note that job centers are not necessarily the peak points of population density.

Step 5. Implementing linear regressions: In preparation for regressions, we need to scale the distance value (r) from meters to kilometers, and calculate the terms $ln(Dr)$, $ln(r)$ and r^2, where r is distance and Dr is corresponding density.

FIGURE 6.2 Population density and job centers in Chicago 2000

Add two Math Formula nodes (Nodes 21 and 22), and link Joiner (Node 13) → Node 21→ Node 22. Configure Nodes 21 and 22 to append two columns dist_sq and lndist with the expressions of $Distance$^2 and ln($Distance$), respectively. They represent *lnr* and r^2.

Add six Linear Regression Learners (Nodes 23–28), and link Math Formula (Node 22) to them. Set the Target and Values as the following pairs in the six nodes, respectively:

```
popden: Distance
popden: lndist
Lnpopden: lndist
Lnpopden: Distance
Lnpopden: dist_sq
Lnpopden: Distance, dist_sq
```

Refer to Table 6.1 in the main book for six regression models. Check the regression results via the output ports of "Coefficients and Statistics", "View: Linear Regression Result View" or "View: Linear Regression ScatterPlot View".

Step 6. Implementing nonlinear and weighted liner regressions: We use Python Script nodes to implement nonlinear and weighted liner regressions with the Python package *scipy* and *statsmodel*.

Add two Python Script nodes 29 and 30, and set up two output ports on Node 30. Copy the Python code in `Nonlinear regression.py` to Node 29, and copy the code in `Weighted OLS.py` to Node 30. Run them and check the results in the output ports.

Refer to Table 6.3 in the main book for all regression results. Both the linear and nonlinear regressions indicate that the exponential function has the best fit among the bivariate models. Newling's model has two explanatory terms and thus is not comparable to the other five models in fitting power.

6.2 PART 2: FUNCTION FITTINGS FOR POLYCENTRIC MODELS AT THE CENSUS TRACT LEVEL

Create a new workflow `Case6B`. Copy the nodes cluster Nodes 1–12 in `Case6A` and paste it to `Case6B`, as they share the same input data. Select Nodes 1–10 and pack them as a component `InputData`, which takes Node 13. Refer to Table 6.2 in the main book for polycentric assumptions and corresponding functions.

Steps 7–9 implement polycentric assumption 1, i.e., only access to the nearest center is necessary for residents. Steps 10–12 implement polycentric assumptions 4, 2 and 3, respectively. Figure 6.3 shows the workflow.

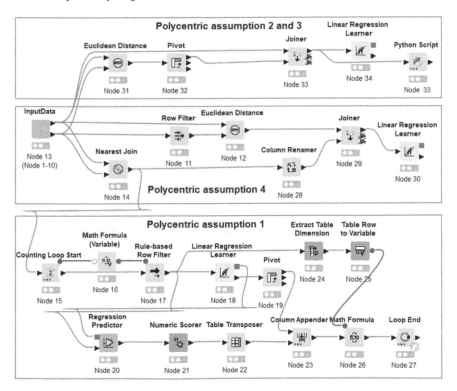

FIGURE 6.3 Workflow of Part 2 of Case Study 6

Step 7. Computing distances between tract centroids and their nearest centers: Add Nearest Join (Node 14), link the component `InputData` to it. In the dialog, check `Inner` for Join mode, and input `100` for Maximum distance, `Kilometer` for Distance unit, and check `Keep CRS from input table`. Click OK and execute it. In the output table, the columns `NID`, `CENT15_` and `Distance` identify each tract centroid, its nearest center, and the distance between them, respectively.

Note the difference between the output tables of Euclidean Distance (Node 12) and Nearest Join (Node 14). The former calculates the distances between census tract centroids and CBD, and the latter yields the distances from census tract centroids to their nearest job centers.

Step 8. Implementing multiple linear regressions on subsets of data: One regression is run on a subset of census tracts sharing the same nearest job center. For 15 centers, there are 15 regressions to implement. This is accomplished by *loop nodes*.

Add Counting Loop Start (Node 15) to the Editor and link Nearest Join to it. In the dialog, set `15` for Number of loops. It creates a flow variable, `currentIteration`, which varies with each time of loop, ranging from 0 to 14. This variable can be used as an intermediate to filter the centers by their IDs.

Add Math Formula (Variable) and link the right flow variable port of Counting Loop Start (Node 15) to it. Configure it to append an integer variable `polyID` with the expression `$${IcurrentIteration}$$+1`.

Add Rule-based Row Filter (Node 17) and link Counting Loop Start (Node 15) to it for consuming data. Link Math Formula (Variable) to the left flow variable port of Node 17 for importing the flow variable `polyID`. In the dialog, input `$CENT15_$ = $${IpolyID}$$ => TRUE` for Expression and keep `Include TRUE matches`. Click OK and execute it to keep the records with `CENT15_=1` (1 for the first iteration of loop).

Add Linear Regression Learner (Node 18) and link Rule-based Row Filter (Node 17) to it. In the dialog, (1) choose `lnpopden` for Target; (2) check `Wildcard/Regex Selection` for Values; (3) input `Distance` for Patterns, check `Widcard`, and make sure that only `Distance` is chosen in the Match (Include) each iteration. Click OK and execute it. Check the result in the output port of Coefficients and Statistics.

We apply pivoting to keep the key result in one row for each regression. Add Pivot[1] (Node 19) and link Linear Regression Learner (Node 18) to it. In the dialog and its Settings tab, (1) under Groups shown in Figure 6.4a, keep empty in Group column(s); (2) under Pivots shown in Figure 6.4b, choose `Variable` in the Pivot column(s); (3) under Manual Aggregation shown in Figure 6.4c, choose `Coeff.` and `P>|t|` for Column, `First` and `First` for Aggregation; under Advanced settings, choose `Pivot name+ Aggregation name` for Column name, and choose `Keep original name(s)` for Aggregation name. Click OK and execute it. The result table contains four columns: `Intercept+Coeff.`, `Intercept+P>|t|`, `Distance+Coeff.` and `Distance+P>|t|`, which are the coefficients and p values for intercept A_i and b_i.

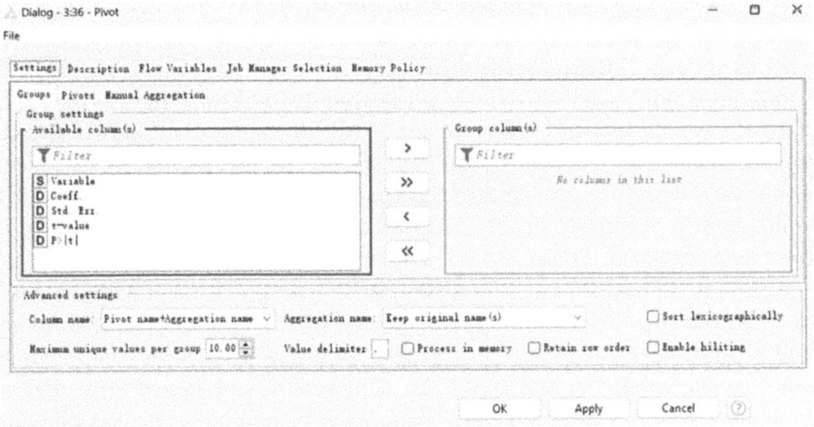

(a)

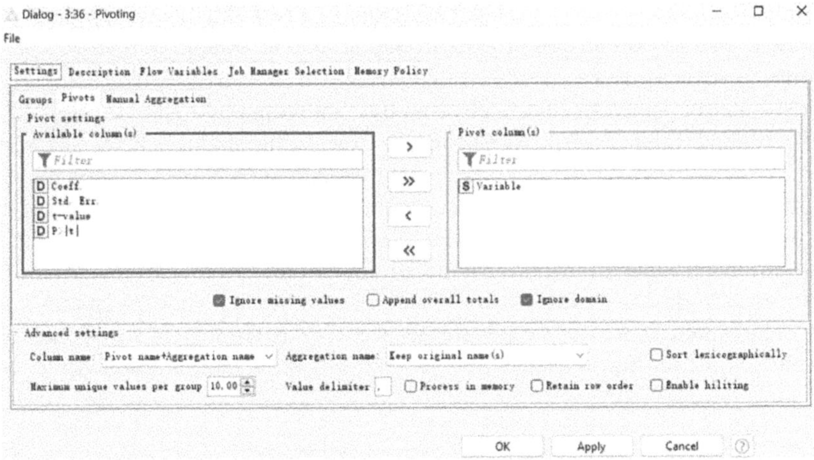

(b)

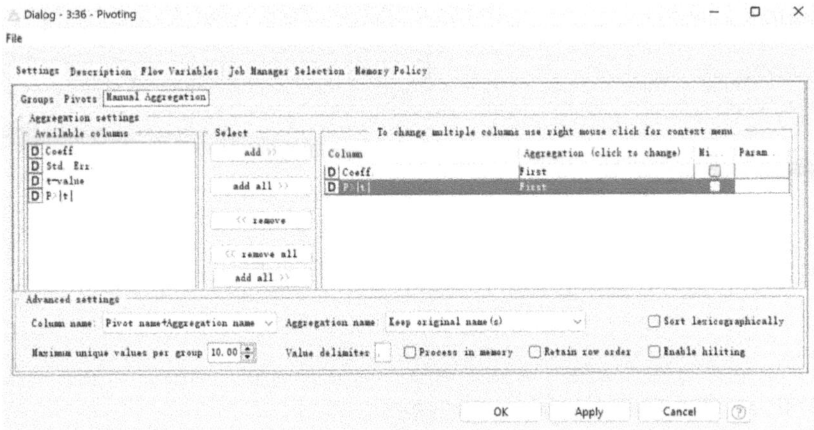

(c)

FIGURE 6.4 Interface of Pivot

Step 9. Calculating the Squared R of linear regression: Linear Regression Learner (Node 18) provides the R^2 value in "View: Linear Regression View", but does not have the port to export the value.

Add Regression Predictor (Node 20), link Rule-based Row Filter (Node 17) to its input data port, and link the output model of Linear Regression Learner (Node 18) to its input model port. Execute it with the default setting (Figure 6.5).[2] It appends a column Prediction (lnpopden) as the predicted value by the model.

Add Numeric Scorer (Node 21) and link Regression Predictor (Node 20) to it. In the dialog shown in Figure 6.6, (1) choose lnpopden for Reference column, Prediction (lnpopden) for Predicted column, (2) under Output column, check Change column name and input Rsquared for Output column name, (3) under Adjusted R Squared, input 1 for Number of predictors,[3] and keep other default settings. Click OK and run it. To combine the coefficient table and score table, we need to select two rows from the score table (n = 7) to correspond with the coefficient table (n = 1).

Add a Table Transposer[4] node (Node 22) and link Numeric Scorer (Node 21) to it. Execute it directly to transpose the entire input table by swapping rows (n = 7) and columns (n = 1).

Add Column Appender (Node 23), link the first output port (Pivot table) of Pivot (Node 19) and Table Transposer (Node 22) to it. In the dialog, check Generate new row RowIDs. Click OK and execute the node to combine the two tables.

If it is desirable to export the number of records for each loop, consider using three nodes: Extract Table Dimension (Node 24), Table Row to Variable (Node 25) and Math Formula (Node 26). Link Rule-based Row Filter (Node 17) → Extract Table Dimension (Node 24) → Table Row to Variable (Node 25). Node 24 creates a table with one column Dimensions with the first row as number of rows, then Node 25 converts it to a flow variable Dimensions. Link Column Appender to Math

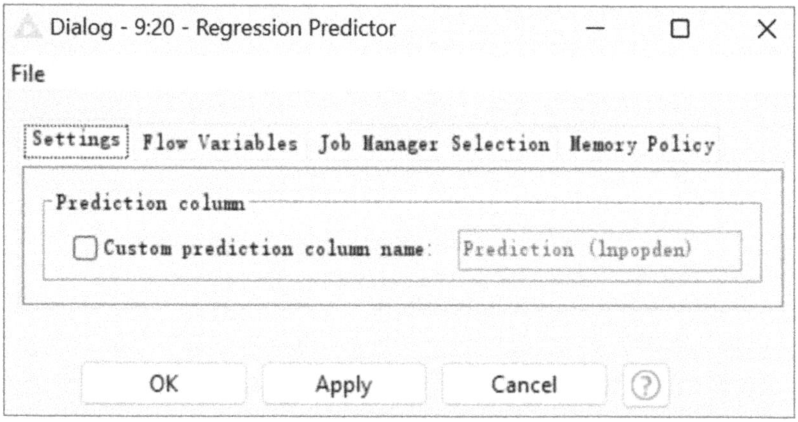

FIGURE 6.5 Interface of Regression Predictor

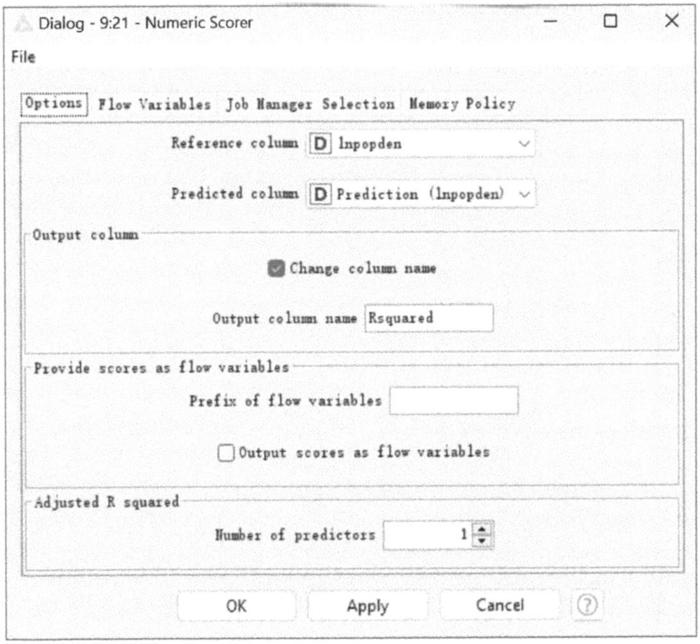

FIGURE 6.6 Interface of Numeric Scorer

Formula (Node 26) and link Table Row to Variable (Node 25) to the flow variable input port of Node 26. Then configure Math Formula (Node 26) to append a column No. tracts with the expression $${IDimensions}$$. Refer to Table 6.4 in the main book for the result.

It is important to note that when using the Table Row to Variable node (Figure 6.7), all selected column names will become variable names, and their corresponding values will be taken from the first row of the table.

Add Loop End (Node 27) and link Math Formula (Node 26) to it. In the dialog shown in Figure 6.8, check Add iteration column, execute it to run the entire loop workflow. The result table records the Coefficients, Statistics and R Squared. Before saving it with CSV Reader. We can use the Round Double node to set the decimal digits for convenience.

Step 10. Exploring polycentric assumption 4: As both Euclidean Distance and Nearest Join use the same column name Distance, we will rename either of them. Add a Column Renamer node (Node 28), and link Nearest Join (Node 14) to it. In the dialog, choose Distance for Column, SubCenterDist for New name. Click OK and execute it.

Add Joiner (Node 29) and link Euclidean Distance (Node 12) and Column Renamer (Node 28) to it. In the dialog of Joiner (Node 29), (1) under Joiner

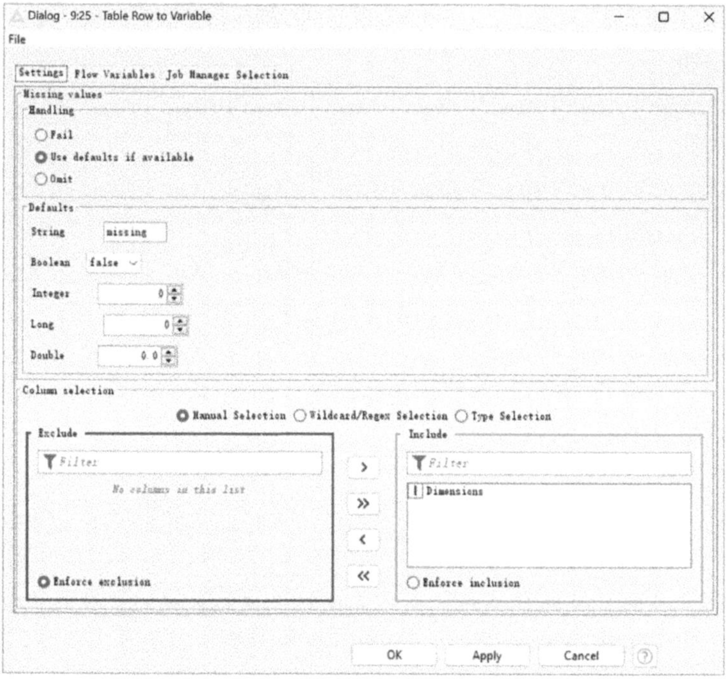

FIGURE 6.7 Interface of Table Row to Variable

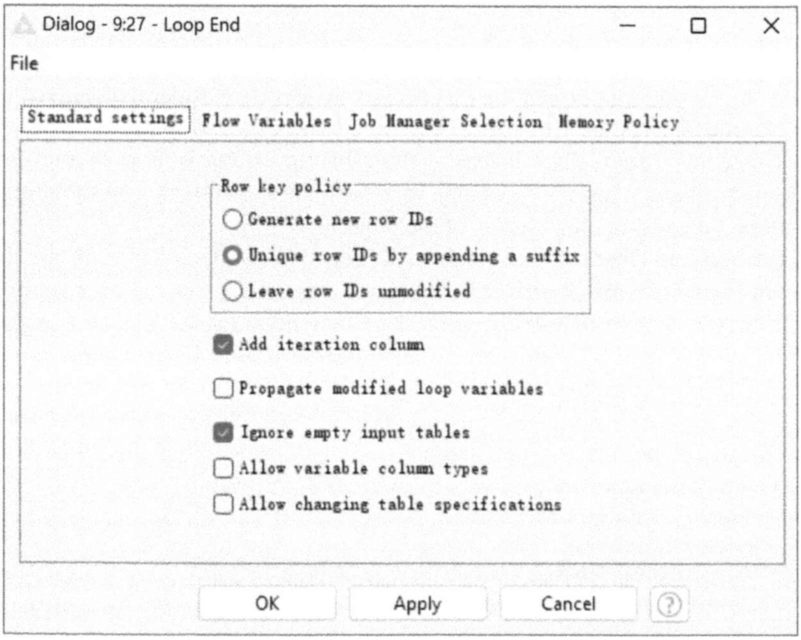

FIGURE 6.8 Interface of Loop End

settings tab, choose `Origin ID` for Top Input ('left' table) and `NID` for Bottom Input ('right' table), and check `making integer types compatible`; (2) under Column Selection, keep all columns in the Include box for both Top Input (left table) and Bottom Input (right table), and keep other default settings. Execute it. This step combines the two distances together: `Distance` for the distance from the CBD and `SubCenterDist` for the distance from its nearest subcenter.

Use Linear Regression Learner (Node 30) to run a regression by setting `lnpopden` as Target and `Distance` and `SubCenterDist` as Values.

The regression model implements polycentric assumption 4, and the result shows how the population density in a tract is jointly influenced by its distance from the CBD and the distance from its nearest subcenter:

$$lnD = 8.914 - 0.0406 \text{ Distance} - 0.0138 \text{ SubCenterDist with } R^2 = 0.442$$

$$(218.23) \ (-31.82) \qquad\qquad (-3.91)$$

The corresponding t-values in parentheses imply that both distances from the CBD and the nearest center are statistically significant, but the distance from the CBD is far more important.

Step 11. Exploring polycentric assumption 2: Copy Euclidean Distance (Node 12) as a new node (Node 31) and link the two output ports of component `InputData` to it. Execute it with the predefined setting to compute distances between tract centroids and all 15 centers.

Add another Pivot (Node 32) and link Euclidean Distance (Node 31) to it for transforming the OD list table to an OD matrix table. In the dialog and its Settings tab, (a) under Groups, keep `Origin ID` in Group columns; (2) under Pivots, choose `Destination ID` in the Pivot column(s); (3) under Manual Aggregation choose `Distance` for Column and `First` for Aggregation; (4) under Advanced settings, choose `Pivot name` for Column name and `Keep original name(s)` for Aggregation name. Click OK and execute it. The result contains 16 columns, `Origin ID` from group column, and 1-15 from pivot column.

Add Joiner (Node 33) and link the top output port of the component `InputData` and the first output port of Pivot (Node 32) to it. In the dialog, (1) under Joiner settings tab, choose `NID` for Top Input ('left' table), and `Origin ID` for Bottom Input ('right' table), and check `making integer types compatible`; (2) under Column Selection, keep all columns in the Include box for both Top Input (left table) and Bottom Input (right table), and keep other default settings. Execute it. This step combines the two distances together: `Distance` for the distances from the CBD, and 1-15 for the distances from the corresponding subcenters.

Add Linear Regression Learner (Node 34) to run a regression. In the dialog, choose `lnpopden` for Target and 1, 2, . . . , 15 for Include box for Values. Click OK and execute it.

This implements the regression model under polycentric assumption 2. Refer to Table 6.4 in the main book for the result. With exceptions of five centers (5, 6, 8 and 12), most of the centers exert influences on regionwide population density patterns as expected with negative coefficients for corresponding distances, and such influences from many centers are significant, e.g., centers 1, 2, 4, 10, 13 and 15.

Step 12. Exploring polycentric assumption 3: For illustration, this step experiments with implementing a model of two centers (center 15 at the CBD and center 5 at the O'Hare airport). The model with all 15 centers contains 30 parameters to estimate and does not converge.

Add Python Script (Node 33), and copy the python script in `nonliner regression assumption 3.py` to run it.

The nonlinear regression result can be written as:

$$D = 10065e^{-0.046 \text{ NEAR\_FID15}} - 6534e^{-0.335 \text{NEAR\_FID5}} \text{ with pseudo-R}^2 = 0.599.$$

$$(3.21) \quad (-2.50) \qquad (1.06) \quad (-1.09)$$

The corresponding t-values in parentheses indicate that the CBD is more significant than the O'Hare airport center.

6.3 PART 3: FUNCTION FITTINGS FOR MONOCENTRIC MODELS AT THE TOWNSHIP LEVEL

The workflow is outlined in Figure 6.9.

Step 13. Using area-based interpolation to estimate population in townships: Create a new workflow `Case6C`. Add Two GeoFile Reader nodes to read the files for census tracts (`trt2k.zip`) and survey townships in the study area (`twnshp.zip`), e.g., E:\KNIME\CMGISV3\Chp06\data\trt2k.zip for Node 1 and E:\KNIME\CMGISV3\Chp06\data\twnshp.zip for Node 2.

Add an Overlay node (Node 3), and link the two GeoFile Reader nodes (Nodes 1 and 2) to it. In the dialog, choose `Intersection` for the overlay mode, check `Return only geometries of the same geometry type`. Click OK and execute the node. In the resulting columns, `AREA_1` and `AREA_2` represent the AREA column in the two input datasets.

Add an Area node (Node 4) and link Overlay (Node 3) to it. Execute it to calculate the area of intersected units.

Add a Math Formula node (Node 5) and link Overlay (Node 3) to it. Configure it to append an integer column `EstPopu` for population with the expression `$POPU$/$AREA_1$* $area$`.

Use a new GroupBy node (Node 6) to summarize the total population by the IDs of townships. In the dialog, (1) under the Groups tab, choose `OBJECTID_2` for Group column(s) and `Keep original name(s)` for Column naming, (2) under Manual Aggregation tab, choose `EstPopu` for Column and `Sum` for Aggregation. Click OK and execute it. Field `EstPopu` in the resulting table represents the estimated population in townships.

Add a Joiner node (Node 7) and link GroupBy (Node 6) and GeoFile Reader (Node 2) to it. In the dialog of Joiner (Node 7), (1) under Joiner Settings tab, choose `OBJECTID_2` for Top Input ('left' table) and `OBJECTID` for Bottom Input ('right' table); (2) under Column Selection, choose `EstPopu` in the Include box for Top Input (left table), keep all

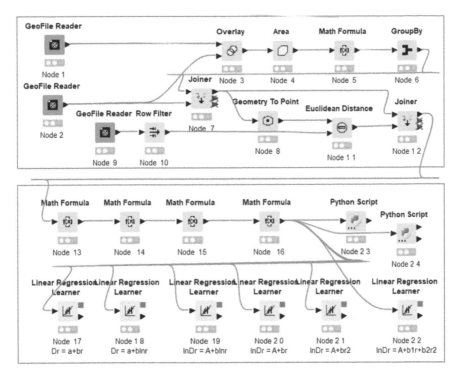

FIGURE 6.9   Workflow of the Part 2 of Case Study 6

columns for Bottom Input (right table), and keep other default settings. Execute it. This step attaches the estimated population to the township data.

Step 14. Computing distances between townships and CBD: First, transform the townships to their centroids using a Geometry to Point node (Node 8). Add a GeoFile Reader node (Node 9) to read the job centers data from E:\KNIME\CMGISV3\ Chp06\data\polycent15.zip. Use a Row Filter node (Node 10) to retain only the CBD with CENT15_ = 15.

Then add a Euclidean Distance node (Node 11), and link Geometry to Point node (Node 8) and Row Filter node (Node 10) to it. In the dialog of Geometry to Point, check centroid for Point Type Selection, Replace for Output Column; in the dialog of Euclidean Distance, choose OBJECTID for both Origin ID and Destination ID columns, and choose Kilometer for Distance unit. Execute them to calculate the distances.

Finally, add a Joiner node (Node 12) to attach the distance values to the township data, and link Joiner (Node 7) and Euclidean Distance (Node 11) to it. In the Joiner dialog, (1) under Joiner Settings, choose OBJECTID for Top Input ('left' table) and Origin ID for Bottom Input ('right' table), (2) under Column Selection, include all columns for Top Input (left table) and choose Distance for Bottom Input (right table) and (3) keep other default setting. Execute it.

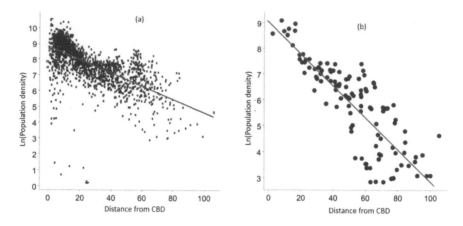

FIGURE 6.10 Density vs. distance with the exponential trend line across (a) census tracts and (b) survey townships in Chicago

Step 15. Fitting monocentric functions with RegressionR: Add two Math Formula nodes (Nodes 13 and 14), and link Joiner (Node 12) → Node 13 → Node 14. Configure the Math Formula nodes to append two columns, `popden` and `lnpopden`, with the expressions `$EstPopu$ * 1000000 / $Shape_Area$` and `ln($popden$)`, respectively.

Copy the node cluster (Nodes 21–30) from `Case6A` and paste it to the Editor. Rename the nodes as Nodes 15–24, and link the Math Formula (Node 14) to Math Formula (Node 15). Execute the workflow to apply the regular linear regression and weighted regression simultaneously. Refer to Table 6.3 in the main book for the result. Given the small sample size of townships, no functions associated with any of the polycentric assumptions are tested.

Figure 6.10a-b shows the fitted exponential function as a straight line after the log-transformation at the census tract and survey township levels, respectively.

6.4 CONCLUDING REMARKS

This chapter practices some basic data processing skills such as linear regression and Euclidean distance calculation, and uses several built-in nodes for various data manipulation tasks. The Column Appender node is used to combine two tables horizontally. The Pivot node is used to transform long tables into wide tables, and here for transforming an Origin-Destination (OD) list (with three columns: Origin ID, Destination ID, and Distance) into an OD matrix (with the Origin ID as the rows and Destination IDs as columns, with corresponding Distance values).

For workflow automation, the Loop End node is used to collect looping results vertically, which is different from the horizontal merging achieved by the Loop End (Column Append) node.

The Extract Table Dimension node is used to extract the number of rows and columns in a table. The Table Row to Variable node is useful for inserting values into other calculations or string manipulation operations such as the Math Formula or Rule-based Row Filter node, as well as String Manipulation or Rule Engine node.

In addition to the commonly used nodes for geospatial analytics such as Euclidean Distance, Spatial Join, Overlay, and Nearest Join, the Kepler.gl Geoview node is used for creating multi-layer visualizations.

NOTES

1 In KNIME AP 5.0, the Pivot node replaces the Pivoting node in KNIME AP 4.7, while the Unpivot node replaces the previous Unpivoting node.

2 Regression Predictor generates a default name for the Target variable in Linear Regression Learner, which can be changed by checking the Custom Prediction column name.

3 The number of predictors is the number of explanatory variables, and *Adjusted-R^2 = 1-(1-R^2)(n-1)/(n-p-1)*, where n is the number of records and p is the number of predictors.

4 In KNIME AP 5.0, the Table Transposer node replaces the Transpose node in KNIME AP 4.7.

7 Principal Components, Factor Analysis and Cluster Analysis and Application in Social Area Analysis

This chapter discusses three important multivariate statistical analysis methods: principal components analysis (PCA), factor analysis (FA) and cluster analysis (CA). PCA and FA are often used for data reduction by structuring many variables into a limited number of components and factors, respectively. The techniques are particularly useful for eliminating variable collinearity and uncovering latent variables. PCA uses the same number of independent components to simply transform the original data, and thus is strictly a mathematical transformation. Users may retain a smaller number of important (principal) components to capture most of the information (variance) of the original complete set of variables. FA uses fewer independent factors to capture most of variance among the original variables with error terms, and thus is a statistical analysis process. While the PCA and FA group variables, the CA classifies observations into clusters according to their attributive homogeneity. In other words, given a data set as a table, the PCA and FA reduce the number of columns and the CA reduces the number of rows.

This case study uses social area analysis to illustrate the application of all three methods. The interpretation of social area analysis results also leads us to a review and comparison of three classic models on urban structure, namely, the concentric zone model, the sector model and the multi-nuclei model. The analysis demonstrates how analytical statistical methods synthesize descriptive urban structure models into one framework. The project is based on a study reported in Gu et al. (2005). The study area was Beijing in 1998, composed of 107 subdistricts (*jiedao*).

The following data sets and Python file are provided under the data folder `Beijing`:

1. feature `subdist.zip` contains 107 urban subdistricts,
2. text file `bjattr.csv` has 14 socioeconomic variables (X1 to X14) for social area analysis, and
3. python code `Factor analysis.py` for factor analysis.

DOI: 10.1201/9781003304357-7

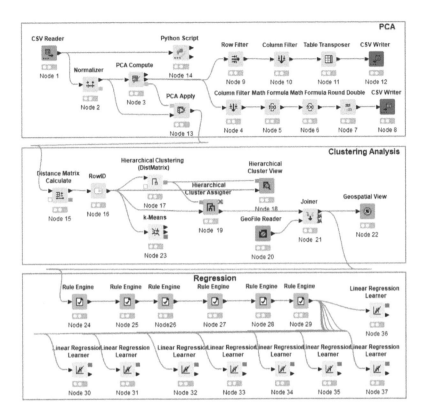

FIGURE 7.1   Workflow of Case Study 7

The attribute table of subdist contains a field sector identifying four sectors (1 for NE, 2 for SE, 3 for SW, and 4 for NW) and another field ring identifying four zones (from the most inner zone 1 to the most outer zone 4). The attribute table of subdist and the text file bjattr.csv contain a common field ref_id identifying the subdistricts.

The study uses principal component analysis (PCA) to derive the constructs (factors) for social area analysis. Factor analysis (FA) is also implemented for illustration but not adopted for subsequent cluster analysis or social area analysis. Figure 7.1 summarizes the workflow of the case study.

7.1   PART 1: PRINCIPAL COMPONENTS ANALYSIS (PCA)

Step 1. Data normalization: First, add a CSV Reader node (Node 1) to import the attribute table bjattr.csv by providing the file path as input, e.g., E:\KNIME\ CMGISV3\Chp07\data\bjattr.csv.

Next, add a Normalizer node (Node 2) and link Node 1 to it. In the dialog shown in Figure 7.2, in the Include box, select all columns except for ref_id, check Min-Max Normalization, and keep the default values of 0 for Min and 1 for Max. Click OK and execute it to apply normalization to all variables.

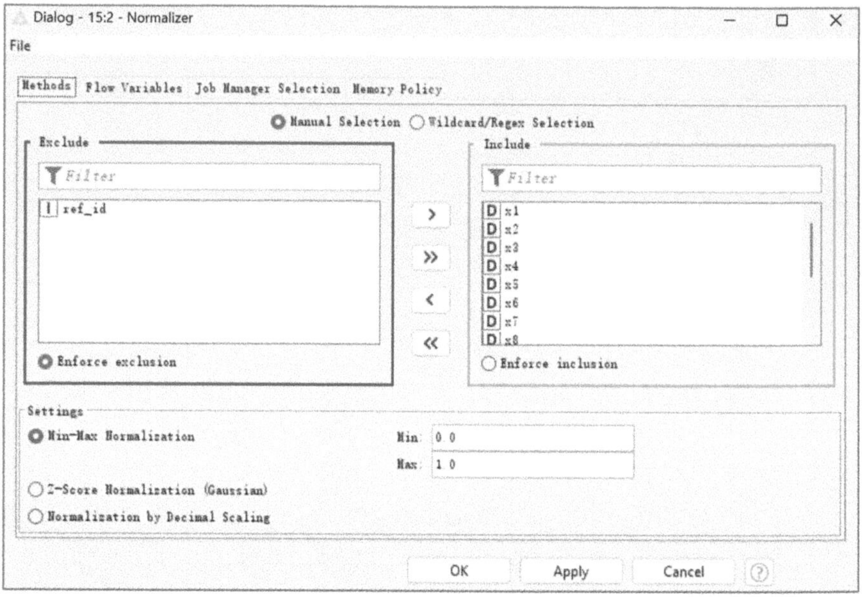

FIGURE 7.2 Interface of Normalization

In the Node Repository, navigate to Analytics > Mining > PCA to access four built-in nodes. To apply principal component analysis, we use the PCA Compute and PCA Apply nodes. Add a PCA Compute node (Node 3) and link Normalizer (Node 2) to it. In the setting, keep all columns but ref_id in the Include box. It contains three outputs: Covariance matrix, Spectral decomposition, and Transformation model.

The Covariance matrix is composed of covariances of each pair of variables. The Spectral decomposition table contains parameters extracted from the PCA, with each row representing one principal component. The rows are sorted with decreasing eigenvalues, each of which represents the variance along the corresponding principal axis. The first column in the table contains the component's eigenvalue, with a higher value indicating higher variance (i.e., the respective component dominates the orientation of the input data). Each subsequent column is labeled with the name of the selected input column and contains a coefficient representing the influence of the respective input dimension on the principal component. The higher the absolute value of the coefficient, the greater the influence of the input dimension on the principal component. The Transformation model holds the PCA transformation used by the PCA Apply node.

Step 2. Eigen value transformation: Various software may use different expressions for the eigenvalues. For instance, the R-based tool used in ArcGIS Pro version is different from the variance generated by PCA Compute, but they share the same algorithm and can be transformed back and forth.

Add a new Column Filter node (Node 4), two Math Formula nodes (Nodes 5 and 6), a Round Double node (Node 7) and a CSV Writer node (Node 8). Link the Spectral decomposition table of PCA Compute (Node 3) → Column Filter (Node 4) → Math Formula (Node 5) → Math Formula (Node 6) → Round Double (Node 7) → CSV Writer (Node 8).

Use the Column Filter node to keep only the column eigenvalue, then use the two Math Formula nodes (Nodes 5 and 6) to append two columns, EGportion and EGtotal, with expressions of $eigenvalue$/COL_SUM($eigenvalue$) and $EGportion$*14, respectively. The Round Double node and CSV Writer node are used to round values and export the table, respectively.

As shown in Table 7.1, the column eigenvalue is extracted directly from the Spectral decomposition table. The value of EGportion represents the ratio of variance captured by each component to the total variance, while the value of EGtotal is obtained by multiplying the proportion of variance explained by the total number of variables (14). Generally, we can select the components with EGtotal > 1, and in this case, the first four as principal components, which account for a cumulative 81.3% (i.e., $0.439 + 0.163 + 0.115 + 0.096 = 0.813$) of the total variance explained.

Step 3. Understanding the principal components analysis: After deciding to choose the four components, we can export the first four rows of the spectral decomposition table. To do so, we can add four new nodes: Row Filter (Node 9), Column Filter (Node 10), Table Transposer (Node 11), and CSV Writer (Node 12). Link the

TABLE 7.1

Eigenvalue Transformation

Principal Component	Eigenvalue	EGportion (Proportion of Variance Explained)	EGtotal (Total Variance Explained)
1	0.261	0.439	6.144
2	0.097	0.163	2.285
3	0.068	0.115	1.604
4	0.057	0.096	1.344
5	0.027	0.046	0.637
6	0.018	0.031	0.428
7	0.016	0.027	0.380
8	0.014	0.023	0.319
9	0.011	0.019	0.268
10	0.008	0.013	0.188
11	0.007	0.011	0.157
12	0.004	0.007	0.103
13	0.004	0.006	0.090
14	0.002	0.004	0.054

Spectral decomposition table of PCA Compute (Node 3)→ Row Filter (Node 9) → Column Filter (Node 10) → Table Transposer (Node 11) → CSV Writer (Node 12).

The Row Filter node and Column Filter node are used to keep the first four rows and filter out the column `eigenvalue`. The resulting table is then transposed by the Table Transposer node and exported by the CSV Writer node. Similar to the standardized loadings (pattern matrix) after the varimax rotation, the factors can be labeled to reflect the major variables loaded:

a. "Land use intensity" is by far the most important factor explaining 43.9% of the total variance and captures mainly six variables: three density measures (population density, public service density, and office and retail density), housing price, and two demographic variables (labor participation ratio and dependency ratio).

b. "Socioeconomic status" accounts for 11.5% of the total variance and includes mostly two variables: average annual income per capita and population natural growth rate.

c. "Ethnicity" accounts for 16.3% of the total variance and includes mostly three variables: ethnic enclave, sex ratio and industry density.

d. "Neighborhood dynamics" accounts for 9.6% of the total variance and includes mostly three variables: floating population ratio, household size and living space.

Step 4. Applying the principal components analysis: To apply principal component analysis, add a PCA Apply node (Node 13) and link the Transformation Model port of PCA Compute node (Node 3) and Normalizer (Node 2) to the corresponding input ports of PCA Apply (Node 13). In the dialog shown in Figure 7.3, (1) check

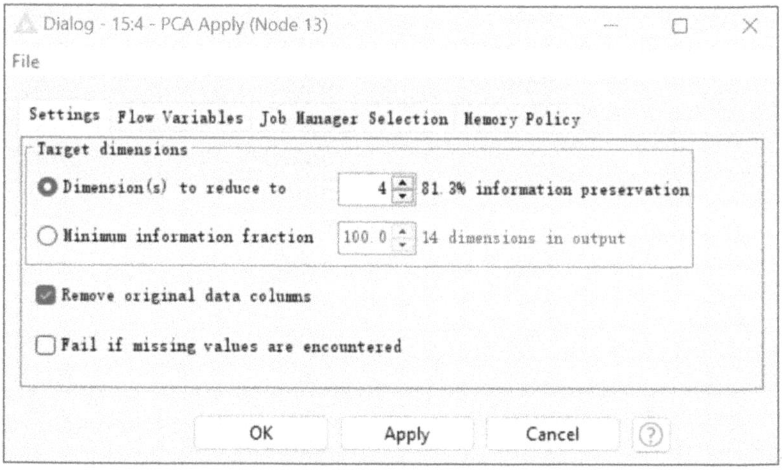

FIGURE 7.3   Interface of PCA Apply

`Dimension(s) to reduce to` and input 4 for it, which indicates the corresponding ratio of information preserved (81.3%),[1] (2) check `Remove original data column`. Click OK and execute it.

The resulting table consists of four columns named as `PCA dimensions 0` (1, 2, 3), which will be used for further clustering analysis.

Step 5. Optional: using Python Script for factor analysis: To apply a Factor Analysis model in KNIME, a Python Script node (Node 14) can be added, configured to have two output ports and linked to the CSV Reader (Node 1). The code from the file `Factor analysis.py` can be inputted in the coding box. The resulting two output tables are similar to the result table generated by the PCA Apply (Node 13) and the factor loading table by the Table Transposer (Node 11), and are identical to the results reported in the main book.

7.2   PART 2: CLUSTER ANALYSIS (CA)

Recall that, a key element in deciding assignment of observations to clusters is *attributive distance*, in contrast to various spatial distance measures discussed in Chapter 2.

Step 6. Calculating Attributive Distance Matrix: Under Analytics > Distance Matrix Calculation, drag Distance Matrix Calculate to add Node 15 and link PCA Apply (Node 13) to the input data port of it. In the dialog shown in Figure 7.4, include all columns except `ref_id` in the Include box, choose `Euclidean` for Distance Selection, keep the default settings for other options, click OK and execute it. Users can explore other distance measurements, such as `Manhattan` and `Cosine`, from the drop-down menu.

To set `ref_id` as the row index and for better illustration, add a RowID node (Node 16) and link Node 15 to it. In the dialog, choose `ref_ids` for New RowID column and check `Remove selected column`. Click OK and execute it.

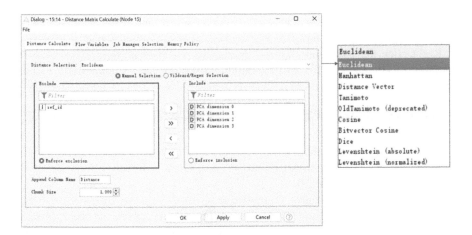

FIGURE 7.4   Interface of Distance Matrix Calculation

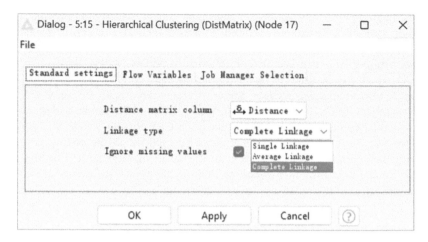

FIGURE 7.5 Interface of Hierarchical Clustering (DistMatrix)

Step 7. Applying Hierarchical Clustering: Under Analytics> Mining> Clustering, drag Hierarchical Clustering (`DistMatrix`) (Node 17) to the Editor and link RowID (Node 16) to it. In the dialog shown in Figure 7.5, choose `Distance` for Distance matrix column, `Complete Linkage` for Linkage type and check `Ignore missing values`. Click OK and execute. It yields a hierarchical cluster tree model that can be fed into the Hierarchical Cluster View and Hierarchical Cluster Assigner nodes.

Add two nodes, Hierarchical Cluster View and Hierarchical Cluster Assigner. Link RowID (Node 16) and the output model of Hierarchical Clustering (DistMatrix) (Node 17) to the corresponding input ports of Hierarchical Cluster View (Node 18) and Hierarchical Cluster Assigner (Node 19). Execute Hierarchical Cluster View (Node 18) directly to view the dendrogram.

In the dialog of Hierarchical Cluster Assigner (Node 19) as shown in Figure 7.6, input `Cluster` for the name, check `Cluster count` and input 9 for Number of clusters to divide the input data into 9 groups based on the distance matrix. Execute it.

Right click the node and choose "Interactive View: Dendrogram" to view the dendrogram (Figure 7.7). To make the output table more concise without the column named `Selected (Hierarchical Cluster Assigner)`, uncheck `Enable selection` under the Interactivity tab in the dialog.

Step 8. Mapping factor patterns and social areas: Add a GeoFile Reader node (Node 20) to read the file `subdist.zip`, e.g., `E:\KNIME\CMGISV3\Chp07\data\subdist.zip`. Add a Joiner node (Node 21), and link the output data port of Hierarchical Cluster Assigner (Node 19) and GeoFile Reader (Node 20) to the input ports of Joiner (Node 21).

In the Joiner (Node 21) dialog, (1) under the Joiner settings tab, choose `Row ID` for Top Input ('left' table) and `Row ID` for Bottom Input ('right' table), and check `string representation` for Compare values in join columns by; (2) under Column Selection, choose all columns in the Include box for both the Top Input (left table) and the Bottom Input (right table), and (3) keep other default settings. Execute the node.

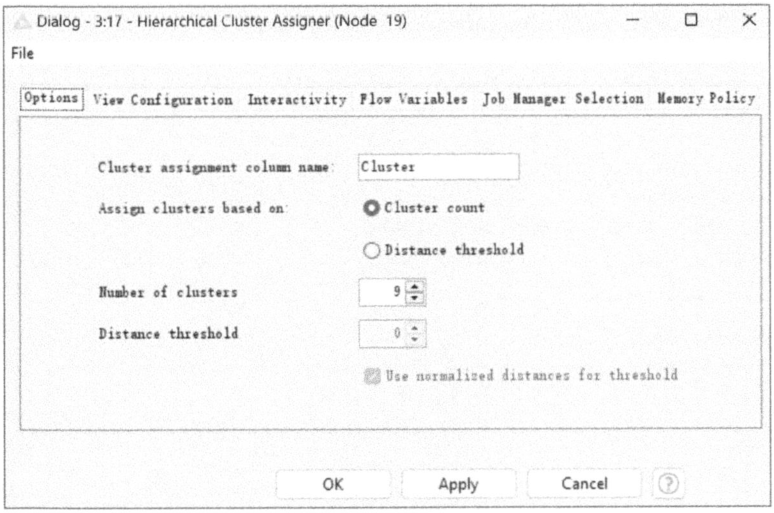

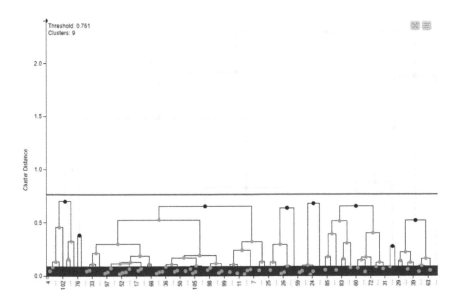

FIGURE 7.6 Interface of Hierarchical Cluster Assigner

FIGURE 7.7   Dendrogram with 9 clusters

Add a new Geospatial View node (Node 22) to visualize the data with Cluster
(or PCA dimension 0-3). Use the columns PCA dimension 0-3 as Marker
color column to map the factor scores of land use intensity, ethnicity, socioeconomic
status, and neighborhood dynamics (refer to Figure 7.8a–d in the main book). Use
the column Cluster as Marker color column to map the nine types of social areas
(refer to Figure 7.9 in the main book).

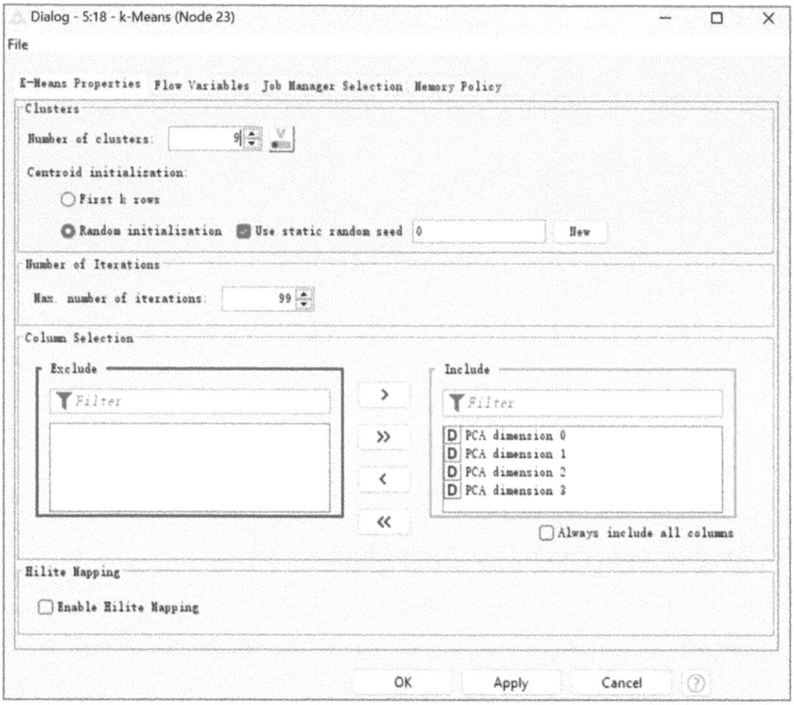

FIGURE 7.8 Interface of K-Means

Step 9. Optional: implementing other clustering methods: KNIME AP provides several built-in tools under the Clustering category. We can test other clustering methods, such as K-Means, K-Medoids, Fuzzy C-Means, and DBSCAN. For example, K-Means is a popular centroid-based clustering algorithm that aims to partition a dataset into K clusters by minimizing the sum of squared distances between data points and their nearest cluster centers.

Add a K-Means node (Node 23) and link RowID (Node 16) to it. In the dialog shown in Figure 7.8, input 9 for Number of clusters, keep all columns in the Include box, keep other default settings, click OK and execute it. The node appends a column `Cluster` to the input table to indicate the grouping result, which can be used for visualization or further analysis.

7.3 PART 3: DETECTING URBAN STRUCTURE MODELS BY REGRESSION

Step 10. Regression with dummy explanatory variables: Two regression models are constructed to test whether the spatial pattern of a factor is better characterized as a zonal or sector model. Based on the circular ring roads, Beijing is divided into four zones from areas in the most inner city (as reference category) towards suburban, coded by three *dummy variables* (x_2, x_3 and x_4). Similarly, three additional dummy

TABLE 7.2

Expression in Rule Engine Nodes

Node	Append Column	Expression	Node	Append Column	Expression
Node 24	x2	$RING$=1=>0	Node 27	y2	$SECTOR$=1=>0
		$RING$=2=>1			$SECTOR$=2=>1
		$RING$=3=>0			$SECTOR$=3=>0
		$RING$=4=>0			$SECTOR$=4=>0
Node 25	x3	$RING$=1=>0	Node 28	y3	$SECTOR$=1=>0
		$RING$=2=>0			$SECTOR$=2=>1
		$RING$=3=>1			$SECTOR$=3=>0
		$RING$=4=>0			$SECTOR$=4=>0
Node 26	x4	$RING$=1=>0	Node 29	y4	$SECTOR$=1=>0
		$RING$=2=>0			$SECTOR$=2=>1
		$RING$=3=>0			$SECTOR$=3=>0
		$RING$=4=>1			$SECTOR$=4=>0

variables (y_2, y_3 and y_4) are used to code four sectors from NE (as reference category) to SE, SW and NW.

Add six Rule Engine nodes (Nodes 24–29) and link them consecutively to Joiner (Node 21), input the following expressions in Table 7.2 to append six columns, x2, x3, x4, y2, y3, and y4 to the input table. A more concise approach for creating dummy variables is to utilize the One to Many node. This method generates columns based on the one-hot encoding of discrete values in a specific column, just like the previous method. However, it also includes an additional column to represent the reference value.

Add eight Linear Regression Learner nodes (Nodes 30–37), and link Rule Engine (Node 29) to them. For the first 4 nodes, set x2, x3, x4 for Values, set PCA dimension 0(1–3) for Target, respectively. For the last 4 nodes, set y2, y3, y4 for Values, set PCA dimension 0(1–3) for Target, respectively. Execute all the notes to run the regressions simultaneously.

The regression results are reported in Table 7.6 of the main book. The land use intensity factor displays a concentric pattern, the neighborhood dynamics pattern is best characterized by the sector model, and the socioeconomic status factor exhibits more of a sector than a concentric structure. However, the ethnicity factor does not conform to either a zonal or sector model, and intuitively resembles a multiple nuclei model.

7.4 CONCLUDING REMARKS

This chapter explores the application of Principal Component Analysis (PCA), Factor Analysis (FA), and Clustering Analysis (CA). A combination of Normalizer, PCA Compute, and PCA Apply nodes are used to implement PCA. We also introduce how to calculate the three forms of Eigenvalues for PCA. FA is implemented using a Python script with the *factor_analyzer* package.

For CA, we emphasize the implementation of Hierarchical Clustering. This involves implementing the Distance Matrix Calculate node and using the Hierarchical Clustering (DistMatrix) node to specify the linkage type, and then specifying the number of clusters in the Hierarchical Cluster Assigner node. This node also provides a dendrogram visualization similar to the Hierarchical Cluster View node. KNIME AP provides many clustering nodes, and the case study also illustrates how to implement the K-Means method.

In preparing data for linear regression, the Rule Engine node plays an important role in assigning multiple values at one time to a single column based on given conditions. The "One to Many" node, which utilizes one-hot encoding, can also serve the same purpose of creating dummy variables with values of 0 and 1.

NOTE

1 Or specify the Minimal information fraction (in percentage) to be preserved.

8 Spatial Statistics and Applications

Spatial statistics analyzes the pattern, process and relationship in spatial (geographic) data. This chapter uses KNIME to implement some fundamental tasks in spatial statistics: measuring geographic distributions, spatial cluster analysis, and spatial regression models.

Section 8.1 uses a case study on place names in Yunnan, China, to illustrate the implementations of various centrographic measures and point-based spatial clustering. Section 8.2 employs a case study to demonstrate the methods for detecting global and local colocation patterns of two types of points (here crimes and facilities). Section 8.3 applies spatial cluster analysis and spatial regression in a case study of homicide patterns in Chicago.

8.1 CASE STUDY 8A: SPATIAL DISTRIBUTION AND CLUSTERS OF PLACE NAMES IN YUNNAN, CHINA

This project is based on a study reported in Wang, Zhang, et al. (2014). Similar to the Case Study 3A on Zhuang place names in Guangxi of China in Chapter 3, this case study also examines place names and shifts the study area to Guangxi's neighboring province, Yunnan. The place names in Guangxi are largely binary and grouped into Zhuang and Han. Yunnan has many ethnic minorities and thus a mix of multiethnic toponyms. Analysis of multiple types of point data presents unique challenges and affords us an opportunity to showcase the values of spatial statistics in historical and cultural geography. The case study has two parts: Part 1 implements the centrographic measures to describe the spatial patterns of toponyms, and Part 2 identifies whether concentrations of one type of toponyms in some areas are random or statistically significant.

Among a total population of 44.5 million in Yunnan, two-thirds are the Chinese majority Han and one-third are ethnic minority groups. More than a quarter of current toponyms in Yunnan are derived from ethnic minority languages. The multiethnic toponyms in Yunnan are grouped into three language branches: Zang–Mian (Tibeto–Burman), Zhuang–Dong (Kam–ai), and Han, all of which belong to the large Sino-Tibetan language family.

Data for the case study is organized in geodatabase `YN.gpkg` under the folder `Yunnan` that includes

1. Three reference layers: `Yunnan_pro` (boundary of Yunnan Province), `Prefectures1994` (Prefectures in Yunnan) and `Majorcity` (Kunming and Dali).

DOI: 10.1201/9781003304357-8

2. Layer `PlaceNames` with (1) field `Large_grou` identifying place names of three language branches (`han`, `zang-mian`, and `zhuang-dong`) and (2) fields `PointX` and `PointY` representing the XY coordinates.
3. Four layers for Han place names corresponding to four different eras: `Han_PreYuan` (before 1279), `Han_Yuan` (1279–1368), `Han_Ming` (1368–1644) and `Han_Qing_postQ` (after 1644).

In addition, the R files for Spatial Scan (i.e., Kulldorff scan in SatScan) are provided in the same folder and include `SatScanR-1.r`, `SatScanR-2.r` and `SatScanR-3.r`.

Figure 8.1 shows the workflow for this case study.

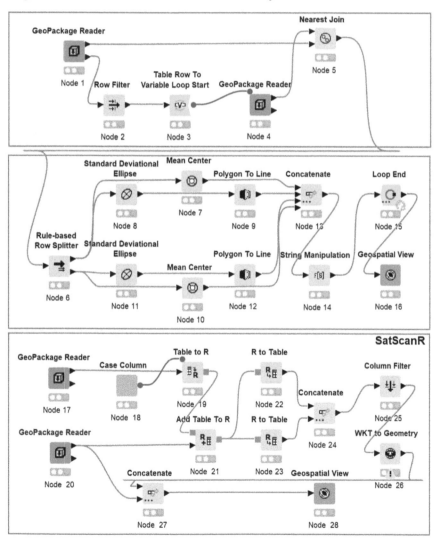

FIGURE 8.1 Workflow of Case Study 8A

8.1.1   PART 1: ANALYSIS OF MULTIETHNIC PLACE NAMES BY CENTROGRAPHIC MEASURES

Toponym entries in the four layers for different eras contain a field `Dynasty` indicating their historical origins, e.g., a value "Ming" in the layer `Han_Ming` means that the place was first named by Han people in the Ming Dynasty and has been preserved till present. This time stamp sheds interesting light on the historical evolution of multiethnic settlement in the region. The centrographic analysis approach is adopted here to reveal the overall spatial trend of historical evolution of Han toponyms. Specifically, as Han gradually moved into the region historically inhabited by native ethnic minorities, it led to *sinification*, a process whereby those minorities came under the influence of dominant Han Chinese. One of the outcomes was that many toponyms of minority language origins got obscured (translated or recorded in Han) or renamed, and it led to a dispersion of Han toponyms over time.

Historically major agriculture development and population settlements were anchored by two cities, Kunming, and Dali (Figure 8.2). Both were major military posts established by the Han rulers to control and exert influence in the

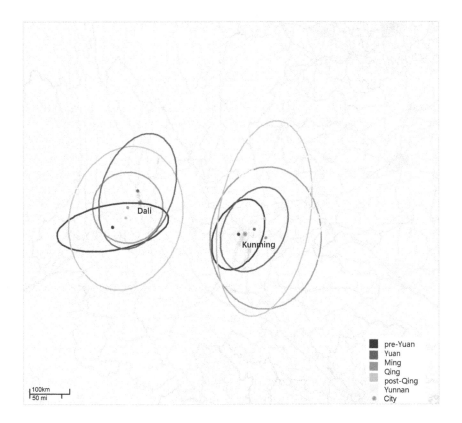

FIGURE 8.2   Mean centers and ellipses of Han toponyms around Dali and Kunming in various eras

region. Distance from whichever of the nearer of the two captures this effect. We begin the analysis by separating the toponyms into two regions, either closer to Dali in the west or closer to Kunming in the east. By doing so, we can examine how the distribution of Han toponyms evolved around the two major cities as springboards of Han's incursion to Yunnan over time. Either subset is then further divided to four layers, corresponding to the Han toponyms named in each of four eras: (1) pre-Yuan (before 1279), (2) Yuan Dynasty (1279–1368), (3) Ming Dynasty (1368–1644), and (4) Qing Dynasty (1644–1912) and post-Qing (after 1912). The centrographic measures are repeated on each of four eras in either of two regions, i.e., eight scenarios.

Step 1. Reading layers with GeoPackage Reader: Create a new workflow named `Case8A`, and add GeoPackage Reader (Node 1) to read the layers from a GeoPackage file, a widely used file format in QGIS. In the dialog box, provide the file path for `YN.gpkg` (e.g., `E:\KNIME\CMGISV3\Chp08\data\YN.gpkg`) and specify the layer name `Majorcity` for Input the layer name or order for reading.[1] The top output table contains the data for the layer `Majorcity`, and the bottom table displays all the layer names that will be loaded into a loop workflow to calculate the mean center and standard deviation ellipse for all four periods.

Add a Row Filter node (Node 2) and connect it to the bottom output table (Geodata layer) of Node 1. In the dialog box, check `Include row by attribute value`, choose `layerlist` for Column to test, check `use pattern matching` and input `Han.*` for it as the pattern, and check `regular expression`. Execute the node to keep the four layer names: `Han_Ming`, `Han_PreYuan`, `Han_Qing_postQ` and `Han_Yuan`.

Add a Table Row To Variable Loop Start node (Node 3) and link it to the Row Filter node (Node 2). Execute the node to generate a flow variable `layerlist` (column name of the input data) with the value of each layer name (row values of the input data) for each time. The interface of the Table Row To Variable Loop Start node bears resemblance to that of the Table Row To Variable node. For reference, please refer to Figure 6.7 in Section 6.2.

Add another GeoPackage Reader node (Node 4). Connect Table Row To Variable Loop Start node to the left variable port of Node 4; and then right-click Node 4, select `Configure Flow Variables . . .` , and set the flow variable `layerlist` for data_layer. Execute the node to read the layer specified by the flow variable `layerlist`.

Add a Nearest Join node (Node 5) and connect the top output port of both GeoPackage Reader nodes (Nodes 4 and 1) to its input ports. In the dialog, choose `Inner` for Join mode, set `1000` for Maximum distance and `Kilometer` for Distance unit, and check `Keep CRS from left input table`. Click OK and execute it. In the result table, the column `NAME` indicates the nearest city for each place.

Add a Rule-based Row Splitter node (Node 6) and link Nearest Join (Node 5) to it. Configure it to split the rows with different city names using the expression `$NAME$="Dali" => TRUE`. The interface of the Rule-based Row Splitter node bears resemblance to that of the Rule-based Row Filter node. For guidance, refer to

Figure 3.3 in Section 3.1. The top output port contains the points having Dali as the nearest city, and the other port contains those having Kunming as the nearest city.

Step 2. Generating the mean center, standard distance and ellipse: Add a Mean Center node (Node 7) and a Standard Deviational Ellipse node (Node 8). Link the top output port of Rule-based Row Splitter (Node 6) to them. Execute Mean Center (Node 7) with the default settings that omit the weight column. It generates the mean center[2] of Han place names in the west region during the Ming dynasty.

Add a new node Polygon To Line (Node 9) and link Standard Deviational Ellipse (Node 8) to it. Execute them directly to generate a standard ellipse and transform it as a line feature for further visualization.

Similarly, copy the cluster (Nodes 7–9) and paste them as Nodes 10–12. Link the top output port of Rule-based Row Splitter (Node 6) to them. Run them to generate the mean centers and ellipses for place names near Kunming.

Add a Concatenate node (Node 13) and configure it to have 4 input ports. Link the output ports of Mean Centers nodes (7 and 10) and Polygon To Line nodes (9 and 12) to it. Execute it to combine the four data (Mean Center, Ellipse for Kunming and Dali) for one period.

To record the layer name, add a String Manipulation node (Node 14) and link Concatenate (Node 13) to it. Configure it to append a new column `Period` with the expression `string($${Slayerlist}$$)`.

Step 3. Running workflow loop and visualizing the data: Add a Loop End node (Node 15) and link String Manipulation (Node 14) to it. In the dialog, uncheck `Add iteration column`, click OK and execute it to run the whole loop workflow for the other three eras. Visualize the result with a new Geospatial View node (Node 16). The output includes 16 layers generated by the Mean Center node and Standard Deviational Ellipse node around the two cities in four periods, as illustrated in Figure 8.2. It reflects the spatial expansion of Han toponyms in Yunnan over time, i.e., from pre-Yuan, Yuan, Ming, to Qing and post-Qing.

8.1.2 PART 2: CLUSTER ANALYSIS OF MULTIETHNIC PLACE NAMES BY SATSCANR

This part of the analysis focuses on the distinctive pattern of toponyms of either minority ethnic group (Zang-Mian or Zhuang-Dong) in comparison to the majority Han by cluster analysis.

Step 4. Importing data with Geopackage Reader: Add a GeoPackage Reader node (Node 17), and input the file path for `YN.gpkg` and then layer name `PlaceNames`.

Step 5. Configuring component for case point adjustment: An adjustable component needs to be created to identify the points with Han placenames as control points and Zang-Mian or Zhuang-Dong placenames as case points.

Add a Single Selection Configuration node (Node 18), and input `casefield` for Parameter/Variable Name, and input `zang-mian` and `zhuang-dong` as two lines for Possible Choice. Click OK and execute to generate the flow variable `casefield` with the default value of `zang-mian`.

Add a Table to R node (Node 19), link the first output port of GeoPackage Reader (Node 17) to Node 19, and connect Single Selection Configuration (Node 18) to the

TABLE 8.1

Code Snippet in Table to R Node (Node 19)

	Code	Explanation
1	library(sf); library(dplyr)	# sf package for processing spatial data; dplyr package for data manipulation and transformation
2	library(spatstat)	# spatstat package for spatial point pattern analysis
3	library(smacpod)	# smacpod package for analyzing spatial point patterns
4	library(maptools)	# maptools package for working with maps
5	nsim <- 49	# Set the simulation time for p-value test
6	controlfield<-"han"	# Set the variable 'controlfield' to "han", as control points
7	casefield<-knime.flow. in[["casefield"]]	# Set the flow variable 'casefield' for adjustable case points
8	all_feature <-st_as_sf(knime.in, wkt = "geometry", crs=4326)	# Convert the input data to a spatial object all_feature using the 'sf' package, specifying the geometry column and coordinate reference system
9	all_feature <-st_transform(all_ feature, 3857)	# Transform the coordinate reference system of the spatial object to projected CRS: EPSG 3857 (Web Mercator)
10	control_feature <-subset(all_ feature, Large_grou==controlfield)	# Subset the spatial object with the value of 'controlfield' in the 'Large_grou' column
11	case_feature <-subset(all_feature, Large_grou==casefield)	# Subset the spatial object with the value of 'casefield' in the 'Large_grou' column

left flow variable port of Node 19. Right-click the Single Selection Configuration node (Node 18) and choose Create Component to wrap it in a new component Case Column (Node 18). Ensure that the dialog of the Component Output node in the component is configured for passing the variable casefield to the downstream workflow.

Step 6. Understanding R codes for Spatial scan test (SatScan): In the dialog of Table to R (Node 19), input the R codes in Table 8.1 (provided in SatScanR-1.r). This node reads the data and splits it into two subsets of control points and case points based on the adjustable flow variable casefield.

Add a GeoPackage Reader node (Node 20), and specify the file path for YN. gpkg and input Yunnan_pro for the layer name. Execute it to read the provincial boundary data.

Add a Table To R node (Node 21), and link the first output port of GeoPackage Reader (Node 20) and the output model port of Table to R (Node 19) to the corresponding input ports of Node 21. In the dialog, input the codes in Table 8.2 (provided in SatScanR-2.r). Execute the node to perform a Kulldorff scan on the case and control points within the boundaries of Yunnan province.

Step 7. Exporting and visualizing data: The resulting data contains two main components, casept and clu_circles, representing the points and circles with

TABLE 8.2
Code Snippet in Table to R Node (Node 21)

	Code	Explanation
1	boundary <- st_transform(st_as_sf (knime.in, wkt = "geometry", crs = 4326), 3857)	# Convert the input data to spatial object with geometry column and specified CRS and transform it to Web Mercator
2	boundary_owin <- as.owin(as_Spatial(boundary))	# Convert the boundary to 'owin' class for use in *spatstat*
3	sf_to_sfpt<-function (inputfeautre, category) {spt <- data.frame(st_coordinates(inputfeautre)) spt$marks <- category return(spt)}	# Define a function to convert a spatial object to a data frame of coordinates with a specified category
4	ctrl_sp <-sf_to_sfpt(control_feature, "control")	# Convert to coordinates with category "control"
5	case_sp <-sf_to_sfpt(case_feature, "case")	# Convert to coordinates with category "case"
6	df <- rbind(ctrl_sp, case_sp)	# Combine control and case
7	pp <- ppp(dfX, dfY, window = boundary_owin, marks = df$marks)	# Create point pattern object (ppp)
8	scan <- spscan.test(pp, nsim = nsim, case = "case")	# Run the spatial scan test
9	clu_circles <- lapply(scan$clusters, function(cluster) {center <- st_ point(cluster$coords[1,], dim = "XY") buffer <- st_buffer(center, dist = cluster$r) sf <- st_as_ sf(data.frame(geom = st_as_text(buffer), loglikrat = cluster$loglikrat, pvalue = cluster $pvalue, centerx = cluster$coords[1, 1], centery = cluster$coords[1, 2], r = cluster$r), wkt = "geom") st_set_crs(sf, st_crs(boundary))})	# Loop over each cluster found by the scan test # Get the center of the cluster # Create a buffer around the center with radius = r # Convert the buffer to an sf object # Set the CRS of the sf object to match that of the boundary
10	clu_circles <- do.call(rbind, clu_circles)	# Combine all circles
11	clu_id <- lapply(seq_along(scan$clusters), function(i) {data.frame(pointID= scan$clusters[[i]]$locids, clusterID = i)})	# Loop over each cluster found by the scan test # Create a data frame with the point IDs and the cluster ID
12	clu_id <- do.call(rbind, clu_id)	# Combine all into one
13	casept <- case_feature %>% mutate (pointID = row_number())%>% left_join(clu_id %>% filter (pointID > nrow(ctrl_sp)) %>% mutate (pointID = pointID—nrow(ctrl_sp)), by = "pointID") %>% st_transform(4326)	# Start with the case features # Add a point ID column # Join with the cluster IDs for the case points # Adjust the point IDs to match the combined data frame # Convert to WGS84 (lat/long)
14	clu_circles <- st_transform(clu_circles, 4326)	# Convert to WGS84 (lat/long)

TABLE 8.3
Code Snippet in R to Table Node (Node 22 and Node 23)

Code	Explanation
Nodes 22	
1 clu_circles1<-st_drop_geometry(clu_circles)	#Drop the geometry column
2 clu_circles1$wkt<-st_as_text(clu_circles$geom)	#Add geometry as wkt column
3 clu_circles1$clusterID<-as.integer(rownames(clu_circles1))	#Create clusterID column by index
4 knime.out <-clu_circles1	#Export to the output port
Node 23	
1 casept1<-st_drop_geometry(casept)	#Drop the geometry column
2 casept1$wkt<-st_as_text(casept$geometry)	#Add geometry as wkt column
3 knime.out <-casept1	#Export to the output port

significant cluster IDs. To export the data, add two R to Table nodes (Nodes 22 and 23), and input the codes shown in Table 8.3 (provided in SatScanR-3.r). Note that the geometry columns are represented by column wkt.

Add a Concatenate node (Node 24) and link the two R to Table nodes to it. In the dialog, check use union columns, and execute it. Add a Column Filter node (Node 25) to keep only the three key columns, wkt, clusterID and pvalue. Add a WKT to Geometry node (Node 26) to transform wkt to geometry column. Then Add another Concatenate node (Node 27) to combine the data from GeoPackage Reader (Node 20) and WKT to Geometry node (Node 26). Add a Geospatial View node (Node 28) to visualize the cluster circles, cluster points, and the base map (set clusterID or pvalue for the Marker color column). A Projection node between Node 20 and Node 27 might be necessary to ensure that the data from Node 20 has the same CRS (e.g., 4326) as the one used in Node 26.

To run the spatial scan for another category, switch the case choice from zang-mian to zhuang-dong in the CasePoints component (refer to step 2), and repeat the workflow.

The case study identifies five toponym clusters for zang-mian (Figure 8.3a) and two toponym clusters for zhuang-dong (Figure 8.3b).[3]

8.2   CASE STUDY 8B: DETECTING COLOCATION BETWEEN CRIME INCIDENTS AND FACILITIES

This case study is based on the work reported in Wang, Hu et al. (2017) and detects colocation between motorcycle thefts and three types of facilities, namely entertainment entities, retail shops and schools, in a city in Jiangsu Province, China. It implements the global and local colocation quotients and explores various ways of defining spatial weights.

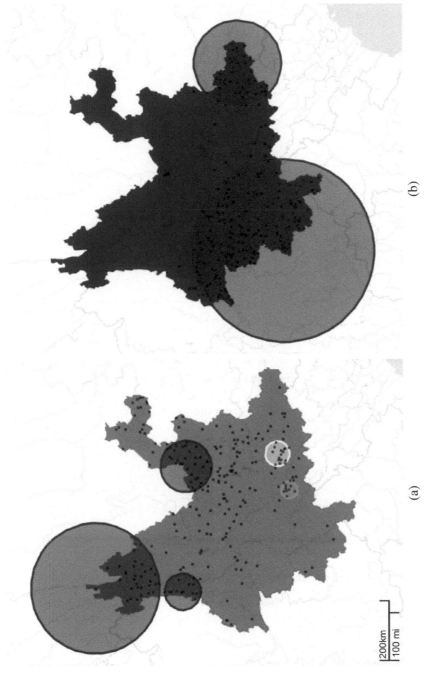

FIGURE 8.3 Minority toponym clusters among Han toponyms: (a) Zang–Mian, (b) Zhuang–Dong

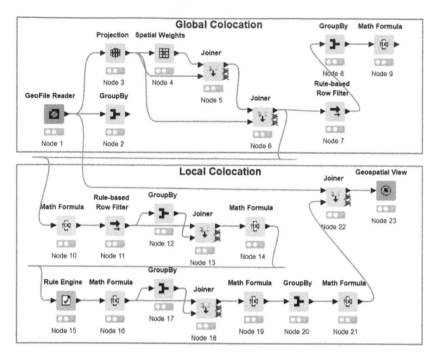

FIGURE 8.4 Workflow of Case Study 8B

Data for the case study is organized in the folder `Jiangsu` and includes:

1. A reference layer `District.zip` for the study area boundary and administrative districts,
2. A network feature dataset `Cityroad.zip` for the road network of the city and
3. A point layer `MotorTheft.zip` with a field `Category`, whose values = `Motorcycle_theft`, `Retail_shop`, `School`, or `Entertainment`, corresponding to four types of points.

Figure 8.4 shows the workflow for this Case Study 8B.

8.2.1 PART 1: GLOBAL COLOCATION QUOTIENT (GCLQ)

Recall that *global colocation quotient* (GCLQ) examines the overall association between observed and expected numbers of type B points in proximity to type A points, formulated as

$$CLQ_{A \to B} = \frac{N_{A \to B} / N_A}{N_B / (N-1)}, \tag{8.1}$$

where N represents the total number of points, N_A denotes the number of type A points, N_B depicts the number of type B points, and $N_{A \to B}$ denotes the number of type A points that have type B points as their nearest neighbors.

In practice, point A may have multiple nearest neighbors within a bandwidth. Under this circumstance, the global CLQ treats each nearest neighbor of point A equally, and $N_{A \to B}$ is formulated as

$$N_{A \to B} = \sum_{i=1}^{N_A} \sum_{j=1}^{nn_i} \frac{f_{ij}}{nn_i}, \tag{8.2}$$

where i denotes each type A point, nn_i represents the number of nearest neighbors of point i, j indicates each of point i's nearest neighbors nn_i, and f_{ij} is a binary variable indicating whether or not point j is of type B under investigation (1 indicates yes and 0 otherwise). Here the notation nn_i, also termed "spatial weights", highlights how the definition of neighbors influences the index.

Step 1. Summarizing ID counts by category of points: Create a workflow Case8B. Add a GeoFile Reader node (Node 1) to the Editor and input the file path of metortheft.zip, e.g., E:\KNIME\CMGISV3\Chp08\data\Jiangsu\metortheft.zip. Use a GroupBy node (Node 2) to summarize the ID counts by the category of points. In the dialog, (1) under the Groups tab, choose Category for Group column(s) and Aggregation method (column name) for Column naming, (2) under Manual Aggregation tab, choose ID for Column and Count for Aggregation. Click OK and execute it. Field Count*(ID) in the resulting table represents the total counts: $N_A = 7777$ (Motorcycle_theft), $N_B = 450$ (Entertainment), and N = 8586 (all four types of points).

Step 2. Applying Spatial Weights to define nearest neighbors: Add a Projection node (Node 3) and link GeoFile Reader (Node 1) to it. In the dialog, input epsg:3857 for New CRS, check Replace for Output column and execute it to transform the data to a projected CRS.

Add a Spatial Weights node (Node 4), and link Projection (Node 3) to it. In the dialog, choose K nearest for Weights category, OBJECTID for ID column, input 10 for Nearest K, keep other default settings, click OK and execute it. This creates a KNN spatial weight list for each point and its 10 nearest neighbors (k = 10), represented by the field focal and neighbor ($n = 8586 \times 10 = 85860$). The field weight indicates that each record shares the same weight, 0.1.

Add two Joiner nodes (Nodes 5 and 6), link Spatial Weights node (Node 4) and Projection (Node 3) to Joiner (Node 5), then link Joiner (Node 5) and Projection (Node 3) to Joiner (Node 6). Use the two Joiner nodes to join the attribute of points to the spatial list by the fields focal and neighbor with the common field OBJECID, respectively, remembering to only include the three key columns X, Y and Category to avoid too many duplicated column names. In the dialog of Joiner (Node 5 or 6), (1) under Joiner Settings tab, choose focal (or neighbor) for Top Input ('left' table) and OBJECTID for Bottom Input ('right' table), and check making integer types compatible; (2) under Column Selection, keep all columns in the Include box for Top Input (left table), choose X, Y and

`Category` for Bottom Input (right table), and keep other default settings. Execute the node.

Step 3. Summarizing spatial weights to calculate GCLQ: The result table of Joiner (Node 6) contains five key columns, `focal`, `neighbor`, `weight`, `Category` and `Category(right)`. Add Rule-base Row Filter (Node 7) and link Joiner (Node 6) to it. In the dialog, input `$Category$=` `"Motorcycle_theft"` `AND` `$Category (right)$` `=` `"Entertainment"` `=>` `TRUE` for the Expression to include the records for points of `Motorcycle_theft` and their neighboring `Entertainment` points (n = 2763).

Add a Group By node (Node 8) to summarize weight. In the dialog, (1) under Groups tab, leave blank for Group column(s) and select `Aggregation method (column name)` for Column naming; (2) under Manual Aggregation, choose `weight` for column and `sum` for Aggregation. Click OK and execute it. The field `Sum(weight)` in the result table indicates $N_{A \to B} = 276.3$, as $N_{A \to B} = \sum_{i=1}^{N_A} \sum_{j=1}^{nn_i} \frac{f_{ij}}{nn_i}$ in Equation 8.2.

Add a Math Formula (Node 9), and link GroupBy (Node 8) to it. Configure it to append a column GCLQ with the expression (`$Sum(weight)$/7777`)/ (450/ (8586−1)), as $CLQ_{A \to B} = \dfrac{N_{A \to B} / N_A}{N_B / (N-1)}$ in Equation 8.1. The result shows that GCLQ = 0.678 for *knn* = 10.

Users can adjust the value for Nearest K in Spatial Weights (Node 4) to test the sensitivity of GLCQ with different neighboring numbers, and revise the expression in Rule-base Row Filter (Node 8) to calibrate the GCLQ values for other category pairs (e.g., motorcycle thefts vs. retail shops, motorcycle thefts vs. schools).

8.2.2 PART 2: LOCAL COLOCATION QUOTIENT (LCLQ)

Recall that the local colocation quotient (LCLQ) is formulated as

$$LCLQ_{A_i \to B} = \frac{N_{A_i \to B}}{N_R / (N-1)} \tag{8.3}$$

$$N_{A_i \to B} = \sum_{j=1(j \neq i)}^{N} \left(\frac{w_{ij} f_{ij}}{\sum_{j=1(j \neq i)}^{N} w_{ij}} \right), \tag{8.4}$$

$$w_{ij} = \exp\left(-0.5 * \frac{d_{ij}^2}{d_{ib}^2} \right) \tag{8.5}$$

where the LCLQ for point A_i relative to type B points has a similar expression as the GCLQ in Equation 8.1, $N_{Ai \rightarrow B}$ denotes the weighted average number of type B points that are the nearest neighbors of point A_i, f_{ij} still represents a binary variable indicating if point j is a marked B point (1 for yes and 0 otherwise), w_{ij} denotes the weight of point j, indicating the importance of point j to the ith A point, d_{ij} is the distance between the ith A point and point j, and d_{ib} denotes the bandwidth distance around the ith A point. Equation 8.5 uses a Gaussian kernel density function to assign spatial weights to each neighbor of point A_i, i.e., a farther neighbor from point A_i corresponds to a smaller weight.

Step 4. Applying Gaussian kernel density function with adaptive distance: Recall that Step 2 kept the coordinates during data joining. Add a Math Formula node (Node 10) and link Joiner (Node 6) to it. Configure it to append a column Dist for d_{ij} in Equation 8.5 with the Euclidean distance expression ((\$X\$-\$X (right)\$)^2+(\$Y\$-\$Y (right)\$)^2)^0.5.

Add a Rule-base Row Filter node (Node 11) and link Math Formula (Node 10) to it. In the dialog, input \$Category\$= "Motorcycle_theft" => TRUE for the Expression to include the records of Motorcycle_theft (n = 7777 × 10 = 77770).

Add a new GroupBy node (Node 12) to summarize the maximum distance to neighboring points by each focal point. In the dialog, (1) under the Groups tab, choose focal for Group column(s) and Aggregation method (column name) for Column naming, (2) under the Manual Aggregation tab, choose Dist for Column and Maximum for Aggregation. Click OK and execute it. Field Max*(Dist) in the resulting table represents the bandwidth distance d_{ib} in Equation 8.5, which is an adaptive bandwidth as it varies for each focal point.

Add a Joiner node (Node 13) and link Rule-base Row Filter node (Node 11) and GroupBy (Node 12) to it. In the dialog of Joiner (Node 13), (1) under Joiner Settings tab, choose focal for both Top Input ('left' table) and Bottom Input ('right' table); (2) under Column Selection, keep all columns in the Include box for Top Input (left table), choose Max*(Dist) for Bottom Input (right table), and keep other default settings. Execute it to attach the bandwidth to the spatial weight list data.

Use a Math Formula node (Node 14) to append a column Wij with the expression exp(-0.5*(\$Dist\$/\$Max*(Dist)\$)^2), which is Equation 8.5.

Step 5. Calculating Local $N_{Ai \rightarrow B}$: Add a Rule Engine node (Node 15) and link Math Formula (Node 14) to it. Configure it to append a column Fij with the following expression in Equation 8.4:

```
$Category (right)$="Entertainment"=>1
NOT $Category (right)$="Entertainment"=>0
```

Use a Math Formula node (Node 16) to append a column WijFij with the expression \$Fij\$*\$Wij\$, which is $w_{ij}f_{ij}$ in Equation 8.4.

Add a GroupBy node (Node 17) to summarize the total value of Wij for each local point by focal. In the dialog, (1) under the Groups tab, choose focal for Group column(s) and Aggregation method (column name) for Column naming, (2) under the Manual Aggregation tab, choose Wij for Column and Sum for

Aggregation. Click **OK** and execute it. The field `Sum(Wij)` in the result table stands for $\sum_{j=1(j\neq i)}^{N} W_{ij}$ in Equation 8.4.

Add a Joiner node (Node 18) and link Math Formula node (Node 16) and GroupBy (Node 17) to it. In the dialog of Joiner, (1) under the Joiner Settings tab, choose `focal` for both Top Input ('left' table) and Bottom Input ('right' table); (2) under Column Selection, keep all columns in the Include box for Top Input (left table), choose `Sum(Wij)` for Bottom Input (right table), and keep other default settings. Execute it to attach the total weight.

Use a new Math Formula node (Node 19) to append a column `wfw` with the expression, `$WijFij$/$Sum(Wij)$`, which is the value for $\dfrac{w_{ij}f_{ij}}{\sum_{j=1(j\neq i)}^{N} w_{ij}}$. Apply another Group By node (Node 20) to summarize the total value of `wfw` for each

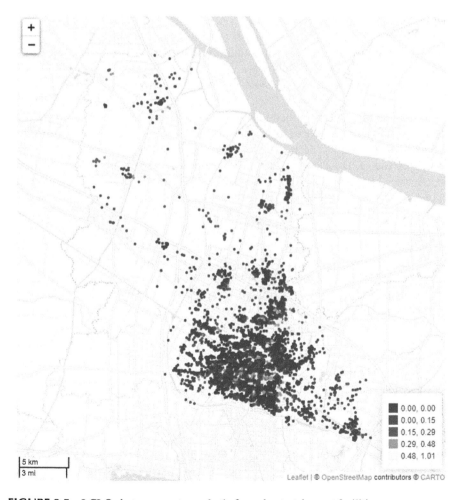

FIGURE 8.5 LCLQs between motorcycle thefts and entertainment facilities

local point by `focal`. In the dialog, (1) under the Groups tab, choose `focal` for Group column(s) and `Aggregation method (column name)` for Column naming, (2) under Manual Aggregation tab, choose `wfw` for Column and `Sum` for Aggregation. Click OK and execute it. The field `Sum(wfw)` in the result table is $N_{A_i \to B}$ in Equation 8.4.

Step 6. Calculating and mapping local colocation quotients: Add a Math Formula node (Node 21) to append a column `LCLQ` with the expression `$Sum(wfw)$/ (7777/(8586-1))`, which represents $LCLQ_{A_i \to B}$ in Equation 8.3.

Add a Joiner node (Node 22) and link GeoFile reader (Node 1) and Math Formula (Node 21) to it. In the dialog of Joiner, (1) under Joiner Settings tab, choose `OBJECTID` for Top Input ('left' table) and `focal` for Bottom Input ('right' table), check `making integer types compatible`; (2) under Column Selection, keep all columns in the Include box for Top Input (left table), choose `LCLQ` for Bottom Input (right table), and keep other default settings. Execute it to attach the local colocation quotient for each point tagged `MotorTheft`.

Add a Geospatial View node (Node 23) to visualize it by setting `LCLQ` as Marker color column (Figure 8.5).

8.3 CASE STUDY 8C: SPATIAL CLUSTER AND REGRESSION ANALYSES OF HOMICIDE PATTERNS IN CHICAGO

The following data sets are provided in the subfolder ChiCity under the study area folder `Chicago`:

1. A polygon feature `citytrt.zip` contains 846 census tracts in the City of Chicago (excluding the O'Hare airport tract) with fields `CNTYBNA` (tract id), `POPU` (population in 1990), `JA` (job accessibility) and `CT89_91` (total homicide counts for 1989–91).
2. A text file `cityattr.csv` contains `cntybna` (tract id) and 10 socioeconomic attributes represented by columns `Field 2-11` based on the 1990 census (families below the *poverty* line, families receiving *public assistance*, *female-headed households*, *unemployment*, *new residents* who moved in the last five years, *renter-occupied* homes, residents with *no high-school diplomas*, *crowded households* with an average of more than 1 person per room, *Black* residents, and *Latino* residents).

Figure 8.6 shows the workflow for Case Study 8C.

8.3.1 PART 1: SPATIAL CLUSTER ANALYSIS OF HOMICIDE RATES

Step 1. Optional: Principal Components Analysis on socioeconomic variables: If needed, refer to Step 1 in Section 7.1. Add a CSV Reader node (Node 1) to read the csv file `cityattr.csv`, e.g., `E:\KNIME\CMGISV3\Chp08\data\Chicago\ cityattr.csv`. Then, use a Normalizer (Node 2) to apply Min-Max Normalization (0–1) for the columns `Field 2-11`.

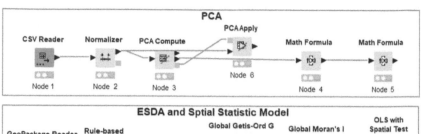

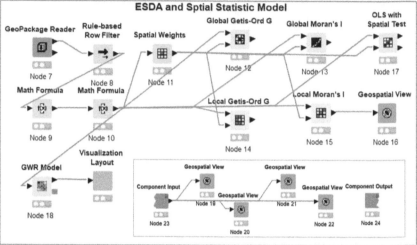

FIGURE 8.6 Workflow of Case Study 8C and sub-workflow in the component Visualization Layout

Add a PCA Compute node (Node 3) and link Normalizer (Node 2) to it. In the dialog, choose `Field 2-11` in the Include box, and execute it. Add two Math Formula nodes (Nodes 4 and 5) to append two columns `EGportion` and `EGtotal`, using the expressions `$eigenvalue$/COL_SUM($eigenvalue$)` and `$EGportion$* 10`, respectively. The eigen values for the first three components in `EGtotal` are 5.73, 2.19 and 0.97, accounting for 57.3%, 21.9% and 9.7% of the total variances (`EGportion`).

Add a PCA Apply node (Node 6) and link the Transformation model (3rd) port of PCA Compute (Node 3) and Normalizer (Node 2) to the corresponding input ports of Node 6. In the dialog, input 3 for Dimension(s) to reduce to, and check `Remove original data column`. Click **OK** and execute it. The three fields `PCA dimension0 (1, 2)` are derived principal components.

Step 2. Computing homicide rates: Add a GeoFile Reader (Node 7) and input the file path for citytract.zip, e.g., `E:\KNIME\CMGISV3\Chp08\data\citytract.zip`. Use a Rule-based Row Filter node (Node 8) to select records of nonzero population with the expression `$POPU$ > 0=> TRUE`. This excludes a tract with no population at the O'Hare airport, and thus avoids zero-denominator in computing homicide rates.

Add a Math Formula node (Node 9) to append a column HomiRate with the expression, $CT89_91$*100000/$POPU$. Use another Math Formula node (Node 10) to append a column LnHomiRate with the expression, ln($HomiRate$+1) (Land et al., 1990, p. 937).[4]

Step 3. Computing Getis-Ord General G and Moran's I: Most nodes in the categories of Spatial Modelling and Exploratory Spatial Data Analysis (ESDA) require spatial weights. Add a Spatial Weights node (Node 11) and link Math Formula (Node 10) to it. In the dialog, choose CITYTRT_ID for ID column, and then execute it directly with the default spatial weight type Queen.

Add a Global Getis-Ord G node (Node 12) and link Math Formula (Node 10) and Spatial Weights (Node 11) to it. Note that the output port of the Spatial Weights node must be connected to the bottom input port of most nodes in Spatial Modelling. In the dialog, choose HomiRate for Variable column and CITYTRT_ID for ID column, click OK and execute it. Right-click on Global Getis-Ord G (Node 12) and choose "Interactive View: Output View" to view the density plot generated by random spatial permutations of the observed attribute values. Since the plot is below conventional significance levels, we reject the null hypothesis of complete spatial randomness in favor of spatial autocorrelation in HomiRate.

The output port of Global Getis-Ord G (Node 12) contains the values for Global Getis-Ord G score, p-value and z-score.

Similarly, add a Global Moran's I node (Node 13) and execute it with the same data link and parameter settings as Global Getis-Ord G (Node 12). The result suggests even stronger clustering, with Moran's I's score as 0.175 and a Z score of 9.274.

Step 4. Mapping local Moran's I_i and local G_i^*: Add a Local Getis-Ord G node (Node 14) and a Local Moran's I node (Node 15), and link Math Formula (Node 10) and Spatial Weights (Node 11) to them. Execute them by choosing HomiRate for the Variable column in both nodes. Local Moran's I (Node 15) is used as an example to examine the association of observations with their neighboring ones. Right-click on it and select "Interactive View: Output View" to view the scatter plot of HomiRate and Spatial Lag of HomiRate.

As shown in Figure 8.7, all the data points are plotted across the four quadrants defined by the mean values. Data points in the upper right (high-high) and lower left (low-low) quadrants indicate positive spatial association. Points in the lower right (high-low) and upper left (low-high) quadrants represent observations that exhibit negative spatial association.

In the output table of Local Moran's I (Node 15), field Moran's I indicates local Moran's I values, and field spots-type indicates five categories of spatial autocorrelation such as HH (High -High), HL (High-Low), LH (Low-High), LL (Low-Low) and Not Significant. Add a Geospatial View node (Node 16) and link Local Moran's I (Node 15) to it, choose spots-type for Marker color column and cubhelix for Color Map, and execute it. See Figure 8.8a for the local Moran's I values and Figure 8.8b for the five types of spatial association.

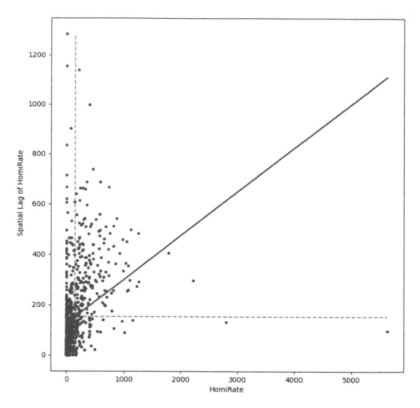

FIGURE 8.7 Moran Scatter Plot

8.3.2 PART 2: REGRESSION ANALYSIS OF HOMICIDE PATTERNS

Part 2 implements several regression models to analyze homicide patterns in Chicago. Three global models are used: the OLS model with spatial test, the spatial lag model (SLM), and the spatial error model (SEM). A local model, the GWR, is employed. As mentioned earlier, the logarithm of homicide rate is used as the dependent variable in all regression models.

Step 5. Implementing the OLS with Spatial Test, SLM and SEM: Add an OLS with Spatial Test node (Node 17), and link Math Formula (Node 10) and Spatial Weights (Node 11) to it. In the dialog, choose LnHomiRate for Dependent variable, FACTOR1, FACTOR2, FACTOR3 and JA for Independent Variables, click OK and execute it. To view the regression report, right-click the node and choose "Interactive View: Model summary view". The report includes the summary of ordinary least squares, regression diagnostic, and diagnostics for spatial dependence, among other things. The output ports contain tables for R Squared, t-value and p-value.

For the Spatial Lag Model (or Spatial Error Model), use the nodes with the corresponding names and define the same parameter settings as the OLS with Spatial Test.

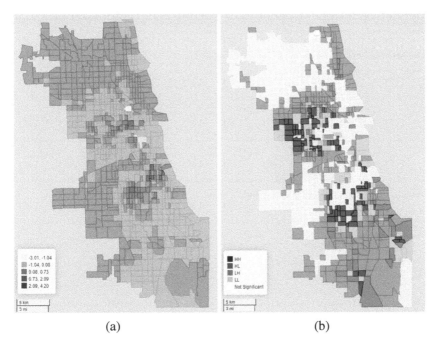

(a)   (b)

FIGURE 8.8   (a) Local Moran's I values, and (b) five types of spatial autocorrelation

Step 6. Implementing the geographically weighted regression (GWR) model: Add a GWR Model node (Node 18) and link Math Formula (Node 10) to it. In the dialog (right panel of Figure 8.9), set LnHomiRate for Dependent variable, FACTOR1, FACTOR2, FACTOR3 and JA for Independent Variables, and keep other default settings (e.g., golden for Search method, 2 for Bandwidth min, bisquare for Kernel).[5] Click OK and execute it.

The regression result consists of three parts: "Interactive View: Model Summary" (left panel of Figure 8.9), Model Coefficients Table (e.g., beta, t-value, and p-value for each data point, which can be utilized for visualization), and Model (for GWR Predictor).

Step 7. Visualization collage: This step uses components to combine multiple visualizations into a single page. Add four Geospatial View nodes and link the output portal of Model Coefficients Table to them. Input FACTOR1_beta, FACTOR2_beta, FACTOR3_beta and JA_beta as Marker color column for each respective node. Then, select all four nodes, right click on any of them, and choose Create component to wrap them into one component Visualization Layout.

Open the component, click the button 🏭 on the top toolbar to open the Node Usage and Layout panel (Figure 8.10). Configure the layout with the templates under Rows, and drag the visualization nodes under Views to the right hand layout. Wherein, we can use + or 🗑 to add or delete columns, and configure ⚙ to set the width and

FIGURE 8.9 Dialog and result of GWR Model node

height for each sub view panel (e.g.,300 for Min and Max Width, 600 for Height as a fix frame). Exit the editing of component, and right click the component to see the combination of visulizations.

As shown in Figure 8.11, the estimated coefficients of factor 1, factor 2, factor 3 and job accessibility (JA) are in fields C1_FACTOR1, C2_FACTOR2, C3_FACTOR3 and C4_JA, and vary across the study area. The coefficients are positive for factor 1 (concentrated disadvantage) and factor 2 (concentrated Latino immigration) and negative for job accessibility, but vary from -0.36 to 0.32 for factor 3 (residential instability).

8.4 CONCLUDING REMARKS

This chapter has three case studies. The first case study uses the GeoPackage Reader node to read *gpkg* files and their layers for further processing within the workflow loop. Mean Center and Standard Deviational Ellipse nodes capture the overall spatial distribution pattern of a set of points in terms of their average

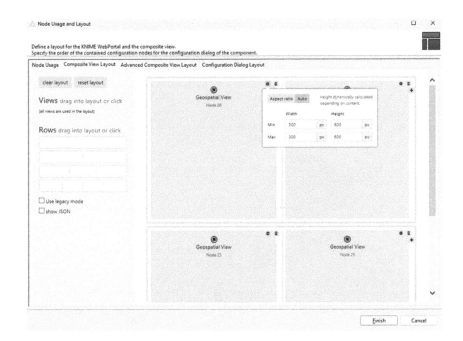

FIGURE 8.10 Dialog of Node Usage and Layout

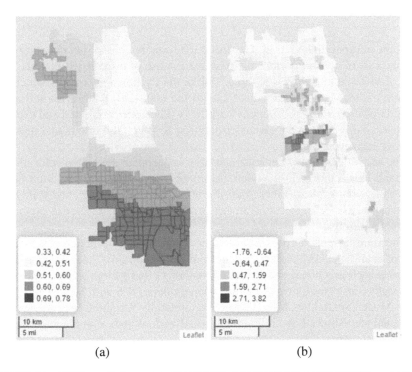

(a)　　　　　　　　　　　　　　(b)

FIGURE 8.11 GWR coefficients of (a) FACTOR1, (b) FACTOR2, (c) FACTOR3, and (d) JA

(Continued)

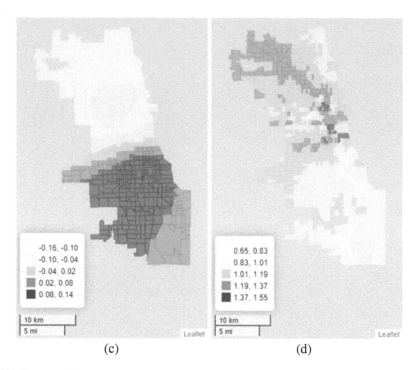

(c)   (d)

FIGURE 8.11 Continued

location and spatial deviation. The Kulldorff scan detects spatial clusters of cases among a set of control points, and requires the use of three R script nodes: Table to R, Add Table to R, and R to table. Unlike the Python Script node, which keeps Python packages and intermediate data in the nodes, all R libraries and data are passed downstream to subsequent workflows. The codes are listed and explained, and users can view the data transfer between R nodes and others and make revisions or improvements.

The second case study demonstrates how to calculate the global colocation quotient and local colocation quotients using only Spatial Weight and built-in nodes such as Rule-based Row Filter, Math Formula, GroupBy, Joiner, and Rule Engine. Users can adjust the k nearest parameter in the Spatial Weights node to explore other parameter settings.

The third case study provides another opportunity to practice principal component analysis (PCA) analysis, and introduces Exploratory Spatial Data Analysis modules such as Global and Local Getis-Ord G and Moran's I. In addition, the case study illustrates four spatial regression models: OLS with Spatial Test, Spatial Error Model, Spatial Lag Model, and Geographically Weighted Regression (GWR). The final step shows how to combine visualizations into a single page using the Component node.

NOTES

1 If unsure about the layer name, input the order (e.g., 0, 1, . . .) or leave the layer name field empty (0 for default) and run the node. Then choose the desired layer name from the bottom output table and rerun the node.

2 It also calculates the standard deviation of the distances between the data points and the mean center.

3 Since the spatial scan statistic is based on Monte Carlo simulations, the results can vary each time you run the analysis due to the random nature of the simulation process. The p-values obtained from the simulations might differ slightly, leading to variation in the clusters detected. Generally, as you increase the number of simulations, the results become more stable and reliable.

4 The logarithmic transformation of homicide rates instead of the raw homicide rates are often used to measure the dependent variable, and 1 is added to the rates to avoid taking logarithm of zero.

5 This setting uses k-nearest as the built-in spatial weights.

9 Regionalization Methods and Application in Analysis of Cancer Data

Rates of rare events (e.g., cancer, AIDS, homicide) often suffer from the *small population (numbers) problem*, where the denominators are small or vary a great deal. The problem can lead to unreliable rate estimates, sensitivity to missing data and other data errors, and data suppression in sparsely populated areas. One approach to mitigate the problem is *regionalization* (or spatial clustering) by grouping smaller areas into larger homogenous regions with comparable population sizes.

The case study in this chapter analyzes variations of breast cancer rates across various constructed regions in Louisiana. Part 1 introduces several one-level regionalization methods such as *SCHC*, *SKATER*, *AZP*, *Max-P* and *REDCAP* methods. Part 2 implements various clustering indicators. Part 3 illustrates the implementation of *Mixed-Level Regionalization (MLR)* method, which decomposes areas of large population and merges areas of small population simultaneously to derive regions with comparable population size. The construction of new regions enables us to map reliable cancer rates.

The case study is developed from the research reported in Mu et al. (2015). The data used in this project is a zip file `LA_Mixtracts.zip` under the folder `Louisiana`, which includes 1,132 census tracts in Louisiana. Its attribute table includes the following fields:

1. `POPU` is population, `Count02` to `Count06` stand for the yearly cancer counts from 2002 to 2006, and `Count02_06` is the 5-year sum of cancer counts across census tracts.
2. `Fact1`, `Fact2` and `Fact3` are three factor scores consolidated from 11 census variables, labeled "Socioeconomic disadvantages", "High health care needs", and "Language barrier" respectively (Wang, Guo, and McLafferty 2012). A higher score of any of the three factors corresponds to a more disadvantaged area. The three factors are used for measuring attribute similarity in the regionalization methods.

The resulting dataset `LA_Result.zip` in Step 4 of Subsection 9.1.2 is also included for user's convenience.

9.1 PART 1: ONE-LEVEL REGIONALIZATION

This section implements several one-level regionalization methods by constructing regions from census tracts. Figure 9.1 shows the workflow of Part 1 of Case Study 9.

DOI: 10.1201/9781003304357-9

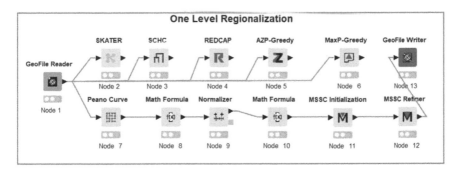

FIGURE 9.1   Workflow of Part 1 of Case Study 9

9.1.1 IMPLEMENTING SKATER, AZP, MAXP, SCHC AND REDCAP METHODS

Step 1. Applying SKATER for regionalization: Create a new workflow Case9A. Add a GeoFile Reader node (Node 1) to the Editor and input the file path LA_ Mixtracts.zip in Node 1 (e.g., E:\KNIME\CMGISV3\Chp09\data\LA_ Mixtracts.zip), and execute the node to read the data.

Under Spatial Clustering, drag the SKATER node to the Editor and link GeoFile Reader (Node 1) to SKATER (Node 2). In the dialog shown in Figure 9.2, choose COUNT06 for Bound column for minibound, FACT1;FACT2;FACT3 for Attribute columns for clustering, 200 for Number of clusters, 16 for Minimum sum value for bound variable in output clusters, and keep Queen checked under Spatial weight mode. Execute it to append a resulting column Cluster ID to the input data. Based on the column Cluster ID, use a GroupBy or Dissolve node to aggregate the result for further analysis.

Step 2. Applying SCHC, REDCAP, AZP-Greedy and MaxP-Greedy: Add four nodes to the Editor, such as SCHC (Node 3), REDCAP (Node 4), AZP-Greedy (Node 5) and MaxP-Greedy (Node 6), and link GeoFile Reader (Node 1) to each of them. These nodes share most of the parameter settings with the SKATER node in step 1. However, SCHC or REDCAP includes an extra option for Linkage model, and MaxP node does not require input for Number of clusters. For more details on these nodes, refer to the Main book.

9.1.2 IMPLEMENTING MSSC AND MPC MODULES IN MLR

This subsection utilizes the Modified Scale-Space Clustering (MSSC) and Modified Peano Curve (MPC) algorithm modules in the Mixed level regionalization (MLR) to perform one-level regionalization.

Step 3. Creating aggregation order by integrating spatial order and attributes: Under Spatial clustering, drag a Peano Curve node (Node 7) to the Editor and link it to the GeoFile Reader (Node 1). Execute it with the default value of 32 for Binary-digit scale. This appends a column peanoorder to indicate the spatial order for each record.

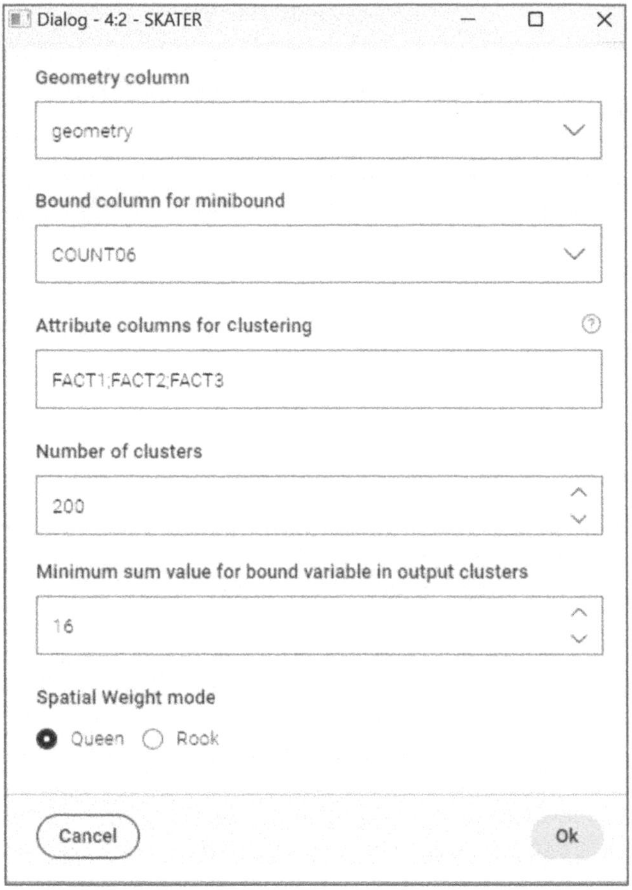

FIGURE 9.2   Interface of SKATER

Add a Math Formula node (Node 8) and link Peano Curve (Node 7) to it. Configure it to append a column `AttriOrd` with the expression `$FACT1$*0.518+$FACT2$*0.1178+$FACT3$*0.0902`, which computes the weighted attributes based on the PCA eigenvalues.

Add a Normalizer node (Node 9) and link Math Formula (Node 8) to it. In the dialog, choose `AttriOrd` in the Include box, keep `Min-Max Normalization` checked and use the default values `0` and `1` for Min and Max. Click OK and execute it to normalize the weighted attributes column `AttriOrd`.

Add another Math Formula node (Node 10) to append a new column `OrdVal` with the expression `$peanoorder$*0.9+$AttriOrd$*0.1`, which integrates the spatial order and weighted attributes with a ratio of `9:1`. Users may experiment with other ratios to see the impacts.

Step 4. Implementing MSSC Initialization and MSSC Refiner: Add a MSSC Initialization node (Node 11) and a MSSC Refiner node (Node 12), and link Math Formula node (Node 10)→ MSSC Initialization (Node 11) → MSSC Refiner (Node 12).

The MSSC Initialization node implements clustering following the ascending aggregated order `OrdVal` with breaking points defined by threshold constraints for cluster columns based on iteration. The MSSC Refiner handles clusters that may not meet the minimum capacity requirements by combining the spatial weight matrix with aggregated order.

In the dialog of MSSC Initialization shown in Figure 9.3a, choose `OrdVal` for Weighted order column, `POPU;COUNT06` for Constraints columns names for clustering, and `20000;16` for Minimum constraint values for clustering, click OK and execute it. The node appends two columns `Cluseter ID` and `isolated`. A further summarization based on the column `Cluseter ID` shows there are 284 clusters generated, of which 172 clusters meet the constraints while 112 clusters do not. The isolate column shows whether a census tract is an isolated unit based on the specific spatial weight mode. The MSSC Initialization node also generates two flow variables from the input parameters, `constraint_capacity` (20000;16) and `constraint_variable` (POPU;COUNT06), which can be used for other MSSC nodes.

In the dialog of MSSC Refiner shown in Figure 9.3b, choose `Cluster ID` for Cluster id column from MSSC, `isolate` for Isolate column from MSSC, `POPU;COUNT06` for Constraints columns names for clustering, and `20000;16` for Minimum constraint values for clustering. Click OK and execute it. Another way to input the constraint column names and values is to use the two flow variables mentioned above. The resulting table in MSSC Refiner node updates the column `Cluster ID`, which shows that all new clusters meet the double constraints of total population and total cancer counts.

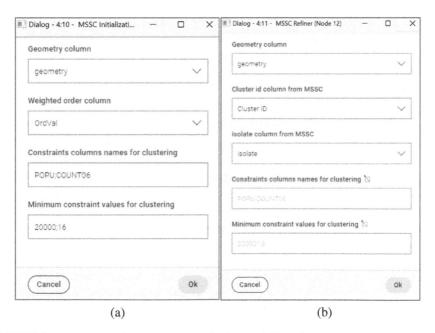

(a)   (b)

FIGURE 9.3   Interface of MSSC: (a) Initialization, and (b) Refiner

The Isolation Tackler node in KNIME is used to handle isolated units in a clustering result. An isolated unit is a unit that is not directly connected to any other units in the cluster based on a spatial weight matrix. The Isolation Tackler node allows users to either remove the isolated units from the clustering result or merge them into nearby clusters based on certain criteria. In this case, since all values in the column isolate are 0, there are no isolated units and thus no need to use the Isolation Tackler node.

Add a GeoFile Writer node (Node 13) and link MSSC Refiner (Node 12) to it. Configure the node to save the data as LA_Result.shp by inputting a proper file path, e.g., E:\KNIME\CMGISV3\Chp09\data\LA_Result.shp.

9.2 PART 2: CALIBRATING CLUSTERING INDICATORS WTVR AND COMPACTNESS

This section demonstrates how to calibrate several indicators for evaluating the regionalization results. *Within-total variance ratio (WTVR)* measures the proportion of variance in attribute data that can be explained by clustering, i.e., total variance within clusters divided by total variance of all data. A higher WTVR indicates better clustering with derived regions that are more homogeneous.

Geometric compactness measures how closely a derived region resembles a geometric shape such as a circle or a square. There are five commonly used measures of compactness. For a region with area A and perimeter P, *Polsby-Popper compactness* = $\dfrac{2\pi\sqrt{A/\pi}}{P}$ and *Schwartzberg compactness* = $\dfrac{4\pi A}{P^2}$. *Reock compactness* and *Convex Hull compactness* are the ratio of the area to the area of minimum bounding circle and convex polygon, respectively. *Length-Width Ratio* is the ratio of the length to the width of the minimum bounding rectangle that surrounds the region. A higher compactness value indicates a more geometrically compact region.

Figure 9.4 shows the workflow of Part 2 of Case Study 9.

9.2.1 CALCULATING WITHIN-TOTAL VARIANCE RATIO (WTVR)

Step 1. Calculating within-cluster variance: As illustrated in Figure 9.5, *within variance* is the sum distance of all points to their respective cluster centroids (A, B, and C), *between variance* is the sum distance of all cluster centroids (A, B, and C) to the overall centroid (P), and *total variance* represents the sum distance of all points to the overall centroid (P). In practice, we can use (x-mean(x))^2 instead of (x-mean(x))^2/(n-1) when calculating within-cluster variance or within-total variance ratio in clustering analysis.

Create a new workflow Case9B. Add a GeoFile Reader node (Node 1) to the Editor and input the file path LA_Result.shp in Node 1 (e.g., E:\KNIME\ CMGISV3\Chp09\data\LA_Result.shp), and execute the node to read the data generated by Step 4 in Subsection 9.1.2.

Add a GroupBy node (Node 2) and link GeoFile Reader (Node 1) to it. Configure it to aggregate the variance of FACT1, FACT2, FACT3 and the count of Rowid_

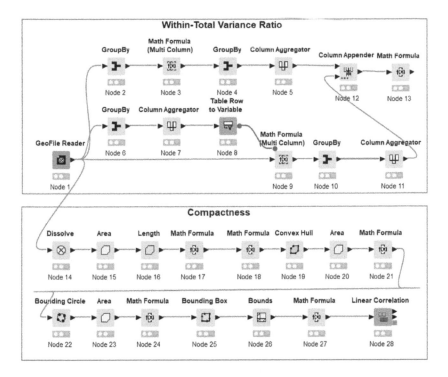

FIGURE 9.4   Workflow of Part 2 of Case Study 9

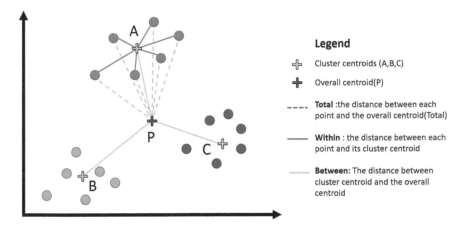

FIGURE 9.5   Illustration of within-cluster, between-cluster and total variances

based on the column Cluster ID, while choosing Aggregation method(-column name) for Column naming. The result is the variance within each cluster based on the built-in function (x-mean(x))2/(n-1).

Add a Math Formula (Multi Column) node (Node 3) and link GroupBy (Node 2) to it. In the dialog shown in Figure 9.6, choose Variance(FACT1), Variance(FACT2), Variance(FACT3) in the Include box, input $$CURRENT_

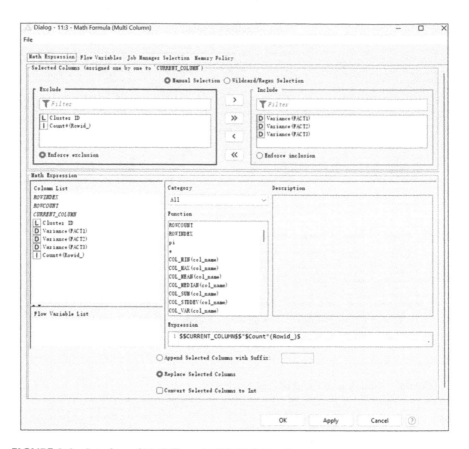

FIGURE 9.6 Interface of Math Formula (Multi Column)

COLUMN$$* ($Count*(Rowid_)$-1) for the Expression, check Replace Selected Columns, click OK and execute it to update and transform three variance columns back to $(x-mean(x))^2$. They can be seen as the distances between all points to their cluster centroids for each cluster in Figure 9.5, e.g., all the blue lines in cluster A.

Use another GroupBy node (Node 4) to aggregate the Sum of Variance(FACT1), Variance(FACT2), and Variance(FACT3), leaving empty for Group column(s) and choosing Keep original name(s) for Column naming. The resulting table contains the total within variance of three variables, which can be seen as the total length of the blue lines in cluster A in Figure 9.5.

Add a Column Aggregator (Node 5) node and link GroupBy (Node 4) to it. In the dialog shown in Figure 9.7, (1) under the Columns tab, keeps all columns in the Aggregation column(s); (2) under the Options tab, choose Sum for Column name and Aggregation method; (3) under Advanced settings, check Remove aggregation columns. Click OK and execute it. The only value Sum is the total within variance for all the three variables.

(a)

(b)

FIGURE 9.7   Interface of Column Aggregator: (a) Columns tab and (b) Options tab

Step 2. Calculating within-total variance ratio: Add a GroupBy node (Node 6) and link GeoFile Reader (Node 1) to it. Configure it to aggregate the Mean of FACT1, FACT2, FACT3, leaving empty for Group column(s) and choosing Keep original name(s) for Column naming. The resulting table contains the mean values for all three variables.

Similarly, use a Column Aggregator node (Node 7) and link GroupBy (Node 6) to it. In the dialog, (1) under the Columns tab, keep all columns in the Aggregation column(s); (2) under the Options tab, choose Mean for Column name and Aggregation method; (3) under Advanced settings, check Remove aggregation columns.

Click OK and execute it. The only value Mean is the mean for all three variables, which can be seen as the total mean Point P in Figure 9.5.

Add a Table Row to Variable node (Node 8) and link Column Aggregator (Node 7) to it. Execute it with the default settings to convert the value Mean as a flow variable.

Add a Math Formula (Multi Column) node (Node 9), link GeoFile Reader (Node 1) to its input data port, and link Table Row to Variable (Node 9) to its input flow variable port. In the dialog, choose FACT1, FACT2, FACT3 in the Include box, input ($$CURRENT_COLUMN$$-$${DMean}$$)^2 for the Expression, and check Replace Selected Columns. Click OK and execute it to update and transform the three columns to their squared deviation $(x-mean(x))^2$. They can be seen as the distances between all points to the overall cluster centroid in Figure 9.5, e.g., all the gray dashed lines in cluster A.

Use a GroupBy node (Node 10) to aggregate the Sum of FACT1, FACT2, FACT3, leaving the Group column(s) empty and choosing Remove aggregation columns for Column naming. The resulting table contains the total variance of three variables.

Similarly, add a Column Aggregator node (Node 11) and link GroupBy (Node 10) to it. In the dialog, (1) under the Columns tab, keep all columns in the Aggregation column(s); (2) under the Options tab, choose Sum for Column name and Aggregation method; (3) under Advanced settings, check Remove aggregation columns. Click OK and execute it. The only value Sum is the total variance for all the three variables.

Add a Column Appender node (Node 12), and link the two Column Aggregator node (Nodes 5 and 11) to it. Execute it to combine the two columns. The final step is to use Math Formula (Node 13) to append a new column ratio with the expression Sum/$Sum(#1)$, which is the *within-total variance ratio (WTVR)*.

9.2.2 CALCULATING COMPACTNESS INDICES

Step 3. Calculating Polsby-Popper and Schwartzberg indices: First, add a Dissolve node (Node 14) and link GeoFile Reader (Node 1) to it. Execute it by setting Cluster ID for Dissolve column to aggregate the tract units according to the MSSC result.

Next, add an Area node (Node 15) and a Length node (Node 16) to calculate the areas and perimeters represented by the column area and length. Then, add two Math Formula nodes (Nodes 17 and 18) to append two columns, Polsby-Popper and Schwartzberg. The expressions for these columns are pi*4*$area$/ ($length$^2) and 2*pi*sqrt($area$/pi) /$length$, respectively.

Step 4. Calculating ConvexHull and Reock indices: To calculate the area of the Convex Hull, use a Convex Hull node (Node 19) to append a new geometry column hull and an Area node (Node 20) to calculate the area of hull. Then use a Math Formula node (Node 21) to append a new column ConvexHull with the expression $area$/$area(#1)$, where $area(#1)$ is the area of the Convex Hull.

TABLE 9.1
Correlation Matrix of Compactness Indicators for MSSC-Derived Regions

	Schwartzberg	Reock	ConvexHull	Length-Width
Polsby-Popper	0.991	0.791	0.538	0.178
Schwartzberg		0.801	0.553	0.204
Reock			0.663	0.177
ConvexHull				0.557

Similarly, add a Bounding Circle node (Node 22) to append a new geometry column `circle`, an Area node (Node 23) to calculate the area of `circle`, and a Math Formula node (Node 24) to append a new column `Reock` with the expression `$area$/$area(#2)$`, where `$area(#2)$` is the area of minimum bounding circle.

Step 5. Calculating Length-Width Ratio: To calculate the Length-Width Ratio, use a Bounding Box node (Node 25) to generate the minimum bounding rectangles with a new geometry column `box`, then use a Bounds node (Node 26) to calculate the bounds coordinates (`minx`, `maxx`, `miny` and `maxy`). Finally, use a Math Formula node (Node 27) to append a new column `Length-Width` with the expression `1/exp(abs(ln($maxy$-$miny$)-ln($maxx$-$minx$)))`, which keeps it as a ratio of the short edge to the long edge with a maximum value of 1.

A Linear Correlation node (Node 28) is used to explore the correlation between these indicators, and the results are summarized in Table 9.1. The Polsby-Popper and Schwartzberg indices are highly correlated, both have a moderately high correlation with the Reock index, and their correlations with Convex Hull or Length-Width Ratio are relatively lower.

9.3   PART 3: MIXED-LEVEL REGIONALIZATION (MLR)

The MLR method defines three scenarios for regionalization for each parish (county) in Louisiana:

(1) Disaggregation (Type I): Both population and cancer count criteria overflow and the values are more than twice the thresholds, so there is a need to break a parish into subregions.
(2) No action (Type II): If a parish meets both criteria, and the value for either criterion is not over twice the threshold, no further action is needed.
(3) Aggregation (Type III): If a parish has at least one unsatisfied criterion, it needs to be aggregated to adjacent parishes to reach the criteria.

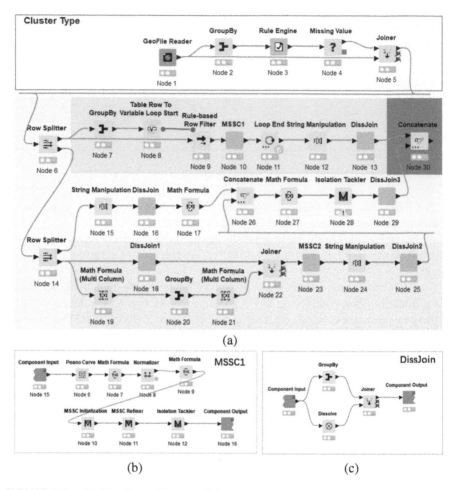

FIGURE 9.8 (a) Workflow of Part 3 of Case Study 9, (b) sub-workflow of MSSC1, and (c) sub-workflow of DissJoin

The KNIME workflow follows this outline as illustrated in Figure 9.8.

9.3.1 IMPLEMENTING MLR ON DISAGGREGATION PARISHES (TYPE I)

Step 1. Identifying three scenarios for parishes: Create a new workflow Case9C. Add a GeoFile Reader node (Node 1) to the Editor and input the file path LA_Mixtracts. zip in Node 1. Execute it to read the data.

Add a GroupBy node (Node 2) and link GeoFile Reader (Node 1) to it. Configure it to aggregate the Sum for POPU and COUNT06 based on the Group column COUNTY, choosing Keep original name(s) for Column naming. The resulting table contains three columns: COUNTY, POPU (population) and COUNT06 (cancer counts) for each parish.

Add a Rule Engine node (Node 3) and link GroupBy (Node 2) to it. Configure it to append a new column TypeIndex with the following expressions, split across two lines:

```
$POPU$ >=40000 AND $COUNT06$ >=32 =>1
$POPU$ < 20000 OR $COUNT06$ <16 =>3
```

This node identifies Type I and Type III parishes, while all other undefined values remain as missing values in the resulting table. Use a Missing Value node (Node 4) to convert all missing values of the data type Number (integer) to 2. This ensures that all three types of parishes are defined by the column TypeIndex.

Add a Joiner node (node 5), and link Missing Value (Node 4) and GeoFile Reader (Node 1) to it. Join the two tables by taking COUNTY for Join columns and only keeping TypeIndex in the Include box of Top Input (left table).

Step 2. Looping for Type I parishes: Add a Row Splitter node (Node 6) and link Joiner (Node 5) to it. In the dialog, check Include rows by attribute value, choose TypeIndex for Column to test, check use pattern matching and input 1 for it. Click OK and execute it to choose the Type I parishes.

Since MSSC will be applied on each Type I parish, a looping workflow based on the county ID is needed. Add a GroupBy node (Node 7) and link the top output port of Row Splitter (Node 6) to it. In the dialog, choose COUNTY for Group column(s), keep other default settings, click OK and execute it to extract the unique FIPS code list of COUNTY.

Add a Table Row To Variable Loop Start node (Node 8) and link GroupBy (Node 7) to it. Execute it directly with default settings to create a flow variable COUNTY whose values use the value of each row and varies with each iteration.

Add a Rule-based Row Filter node (Node 9), link Table Row To Variable Loop Start (Node 8) to the input flow variable port of Node 9, and link the top output port of Row Splitter (Node 6) to the input data port of Node 9. In the dialog, check Include TRUE matches and input $COUNTY$=$${SCOUNTY}$$=> TRUE for the Expression, click OK and execute it to only keep parishes with the same FIPS code in COUNTY.

Step 3. Iterating MLR method on Type I parishes: To implement the same workflow for MSSC and MPC as Part 1 in Section 9.1, we can select the node cluster for MSSC, i.e., Nodes 7–12 in the workflow Case9A, right-click on any node, and choose Create Component to wrap them up as a component MSSC1. Configure MSSC1 to have one input port and one output port. Copy the component MSSC1 and paste it into Case9C. In the Editor of Case9C, link Rule-based Row Filter (Node 9) to the component MSSC1 (Node 10).

Right click the component MSSC1 and choose Open to enter the sub-workflow in the component (Figure 9.8b). Add an Isolation Tackler node (Node 13) and link MSSC Refiner (Node 12) to it while using the same parameters of MSSC Refiner in the Isolation Tackler node's dialog. Remember to link an Isolation Tackler node to the Component Output node in the component. This Isolation Tackler node is necessary as there might be some enclaved census tracts in parishes that have no neighboring units. Exit the editing of the component by clicking the workflow of

Case9C. Execute the component MSSC1 (Node 10) to run the MPC and MSSC on the first parish.

Add a Loop End node (Node 11) and link the component MSSC1 to it. In the dialog, only check Unique row IDs by appending a suffix, Ignore empty input tables, and Allow changing table specifications, click OK and execute it to run the loop. The result table has two new columns, isolate and Cluster ID, similar to Step 4 in Part 1.

Step 4. Aggregating census tracts in Type I parishes: Users may notice duplicated values in the column Cluster ID, as they represent the cluster number in each county and need to be further processed.

Use a String Manipulation node (Node 12) to append a new column FiClus that combines the parish type, FIPs code, and cluster ID of each census tract in each parish as a unique cluster ID with the expression string($TypeIndex$) +"."+string($COUNTY$)+"."+ string($Cluster ID$).

Add three nodes—GroupBy, Dissolve, and Joiner—and wrap them up as a component DissJoin (Node 13) with one input port and one output port to aggregate the attribute and geometry of each cluster. Link String Manipulation (Node 12) to the component DissJoin (Node 13).

Within the component DissJoin (Figure 9.8c), link the Component Input to the GroupBy node and the Dissolve node separately, then link the two nodes to the Joiner node, and finally link the first output port of the Joiner node to the Component Output node.

In the GroupBy dialog, choose FiClus as Group columns, Sum for COUNTY06 and POPU as aggregation, and select Keep original name(s) for Column naming. In the Dissolve dialog, choose FiClus for Dissolve column. In the Joiner dialog, choose FiClus for Join columns, and keep only one FiClus in the Column Selection.

Execute the component DissJoin. The resulting table contains four columns, FiClus, COUNT06, POPU and geometry, which represent the unique cluster ID, aggregated cancer counts, population and geometry, respectively.

9.3.2 Implementing MLR on No-Action Parishes (Type II)

Step 5. Aggregating census tracts in no-action parishes: For Type II parishes, we only need to create a compatible cluster ID and perform aggregation. First, add another Row Splitter node (Node 14) and connect the bottom output port of Row Splitter (Node 6) to it. Use the same configuration as Row Splitter (Node 6) except inputting 2 for pattern matching. Click OK and execute it to split the Type II and III parishes.

Use a String Manipulation node (Node 15) to append a new column FiClus with the expression string($TypeIndex$)+"."+string($COUNTY$)+"."+ string(0), which combine the parish type, FIPs code and 0 as a new column FiClus.

Copy the component DissJoin (Node 13) and paste it as a new component (Node 16), then link String Manipulation node (Node 15) to it. Execute it directly to aggregate all census tracts in each Type II parish as clusters. It shares the same column names with the previous `DissJoin` component.

Since the MSSC will be applied to Type III parishes with an additional column `isolate`, we need to build a column for Type II parishes as well. Add a Math Formula node (Node 17) and link the component DissJoin (Node 16) to it. Configure it to append a new column `isolate` with the expression 0, which means no isolated units in the clusters.

9.3.3 IMPLEMENTING MLR ON AGGREGATION PARISHES (TYPE III)

For Type III parishes, we will first aggregate the attributes of population, cancer counts and clustering attributes from census tracts to the county level, and then implement the regionalization algorithm based on the aggregated attributes.

Step 6. Aggregating census tracts in Type III parishes: To aggregate the attributes of population, cancer counts and geometry, copy the component `DissJoin` (Node 13 or 16) and paste it as a new component `DissJoin1` (Node 18), and link the bottom output port of Row Splitter node (Node 14) to it. In the sub-workflow of component `DissJoin1`, choose COUNTY for both Group columns in the dialog of GroupBy node and Dissolve column in Dissolve node, and keep other predefined settings, then execute the component to aggregate population, cancer counts and geometry.

To aggregate the clustering attributes, add a Math Formula (Multi Column) node (Node 19) and link the bottom output port of Row Splitter node (Node 14) to it. In the dialog, choose FACT1, FACT2 and FACT3 in the Include box, input $$CURRENT_COLUMN$$*$POPU$ in the Expression, and check Replace Selected Columns. Click OK and execute it to weight the three clustering attributes with their respective population.

Add a GroupBy node (Node 20) and link Math Formula (Multi Column) (Node 19) to it. Configure it to aggregate the Sum of FACT1, FACT2, FACT3 and POPU based on column COUNTY, and choose Keep original name(s) for Column naming. The result is sum of population-weighted attributes and total population for each parish.

Add another Math Formula (Multi Column) node (Node 21) and link GroupBy (Node 20) to it. In the dialog, select FACT1, FACT2 and FACT3 in the Include box, input $$CURRENT_COLUMN$$/$POPU$ in the Expression box, check Replace Selected Columns, click OK and execute it to calculate population-adjusted attributes for parishes.

Add a Joiner node (Node 22) and link the component DissJoin1 (Node 18) and Math Formula (Multi Column) (Node 21) to it. Configure it to join the two tables based on the COUNTY column, and remember to exclude the duplicated columns such as COUNTY and POPU from either the top or bottom table in the Column Selection tab.

Step 7. Applying MLR on Type III parishes: Copy the component MSSC1 (Node 10) and paste it as a new component MSSC2 (Node 23), and link Joiner (Node 22) to it. In the component MSSC2 (Node 23), delete the Isolation Tackler (Node 12) and link MSSC Refiner to Component Output, then exit the editing. Execute the component MSSC2 (Node 23) to run MLR. In the resulting column isolate, two parishes are marked as isolated and represented by the value 1.

Add a String Manipulation node (Node 24) to append a new column FiClus with the expression string("3")+"."+"000"+"."+ string($Cluster ID$). Similar to Type II parishes, we aggregate these Type III parishes based on FiClus. Copy any component DissJoin as a new component DissJoin2 (Node 25) and link String Manipulation (Node 24) to it. In the sub-workflow of the component DissJoin2, revise the settings of the GroupBy node to aggregate the Sum for COUNT06 and POPU, and Minimum for isolate based on the column FiClus, and choose FiClus for Dissolve column in the Dissolve node. Execute it to aggregate Type III parishes.

9.3.4 Aggerating All Clusters as MLR

Step 8. Implementing Isolation Tackler on aggregated Type III and II parishes: Add a Concatenate node (node 26) and link Math Formula (Node 17) and DissJoin2 (Node 25) to it. Execute it with the default settings to combine the aggregated Type II and III parishes.

Add a Math Formula node (Node 27) to append a new integer column Cluster ID with the expression $$ROWINDEX$$+1, which assigns a unique ID to each row based on its index. Then add an Isolation Tackler node (Node 28) and link Math Formula (Node 27) to it. Configure its parameters the same as for MSSC Refiner or Isolation Tackler in the component MSSC1 (Step 3 in Part 3). Execute it to update the values in the column Cluster ID, which shows that two isolated Type III parishes are merged to two Type II parishes, respectively.

Step 9. Concatenating all parishes to conclude MLR: Copy any component DissJoin as a new component DissJoin3 (Node 29) and link Isolation Tackler (Node 28) to it. In the sub-workflow of DissJoin3, revise the setting of the GroupBy node to aggregate the Sum for COUNT06 and POPU, First for FiClus based on the column Cluster ID; and choose Cluster ID for Dissolve column in the Dissolve node. Execute it to aggregate Type II and III parishes.

Add a Concatenate node (Node 30) and link the component DissJoin (Node 13) and DissJoin3 (Node 29) to it. Execute the Concatenate node to combine the results for all three types of parishes as the final MLR model output.

9.4 CONCLUDING REMARKS

This chapter first implements five popular regionalization methods, namely SCHC, SKATER, REDCAP, AZP, and Max-P. All incorporate a threshold population constraint to construct regions of sufficient size that are made of one-level areas. Then the chapter focuses on the implementation of MLR that creates regions of comparable population size but made of different levels of area units, such as census tracts or whole counties. The core modules of MLR are MSSC and MPC, which are also applied to one-level areas to illustrate the clustering process.

Two types of indicators for evaluating the regionalization results are introduced: within-total variance ratio for clustering attributes and geometric compactness measures such as Polsby-Popper, Schwartzberg, Reock, Convex Hull, and Length-Width Ratio.

Several new functions are introduced to demonstrate the power of components in KNIME AP. Loop workflows play an important role in applying MLR to Type I parishes so that the same clustering process is repeated on each parish to construct regions while preserving the parish boundaries. Additionally, the nodes in the Geospatial Analytics Extension, such as Spatial Calculation, Bounding Circle, Bounding Box, Bounds, and Convex Hull, are used to calculate compactness indicators.

10 System of Linear Equations and Application of Garin-Lowry Model in Simulating Urban Population and Employment Patterns

This chapter introduces the method for solving a *system of linear equations (SLE)*. The method is fundamental in numerical analysis (NA) and often used as a building block in other NA tasks such as solving a system of nonlinear equations and the eigenvalue problem. Here, the SLE is illustrated in the *Garin-Lowry model*, a model widely used by urban planners and geographers for analyzing urban land use structure. In short, given a basic employment pattern and a distance matrix, the Garin-Lowry model derives the population and service employment patterns.

The case study in this chapter uses a hypothetical city to illustrate how the distributions of population and employment interact with each other as formulated by the Garin-Lowry model, and how the patterns can be affected by the initiation of a basic employment pattern, changing travel friction coefficients and various transportation networks.

10.1 IMPLEMENTING THE GARIN-LOWRY MODEL IN A HYPOTHETICAL CITY

The Case Study 10 uses a hypothetical city to illustrate the Garin-Lowry model. Various scenarios are simulated to help us understand some empirical observations as reported in the literature. The case study is built upon the work reported in Wang (1998).

As shown in Figure 10.3 of the main book, the city is partitioned by a transportation network made of 10 circular rings and 15 radial roads. The CBD tract as a circle occupies a central location, and nine concentric rings expand to the edge with an equal width and intersect with 15 sectors to form a total of $1 + 9*15 = 136$ tracts.

The data folder `SimuCity` contains:

1. a polygon feature `tract.zip` (136 tracts) and its centroids feature `trtpt.zip`,
2. a feature dataset `road.zip` for the road network, and

                                         DOI: 10.1201/9781003304357-10

3. Python files for matrix calculation include `I-GTmatrix.py` for I-GT, `Matrixdot.py` for matrix dot product, and `SLE.py` for solving systems of linear equations.

Features `tract` and `trtpt` contain similar attribute fields: `AREA` (square kilometer), `PERIMETER` (kilometer), and `BEMP_CBD` and `BEMP_Unif` for predefined basic employment in the basic case and the uniform distribution case, respectively (see steps 4 and 5). The two features also save the results calibrated in the four scenarios such as basic case, uniform distribution case, $\alpha = \beta = 2.0$ and the case with the beltway: `POP_CBD`, `POP_Unif`, `POP_A2B2` and `POP_Belt` for population, and `SEMP_CBD`, `SEMP_Unif`, `SEMP_A2B2` and `SEMP_Belt` for service employment. The road network feature `road` contains two fields, `speed` and `speed1`. Field `speed` is a uniform speed (30 mph) on all road segments, and another field `speed1` defines the travel speeds after the beltway is constructed in the 7th ring (defined by the value `r7` in the column `TYPE`). Note that the speed on that beltway is assumed to be 2.5 times (e.g., 75 mph) of the speed on others (30 mph), and thus its impedance in travel time is 1/2.5 of others.

The focus of the case study is to examine the effects of (1) exogenous variations in the basic employment pattern, (2) changes in the travel friction parameters α and β, and (3) construction of a suburban beltway. In all scenarios, total basic employment is $B_T = 100$ with parameters $h = 2.0$ and $e = 0.3$. Denoting total service employment as S_T and total population as P_T, we have: $S_T = 0.3P_T$ and $P_T = 2E_T = 2(B_T + S_T) = 2(100 + S_T)$. Solutions to P_T and S_T are 500 and 150, respectively.

Figure 10.1 shows the workflow of Case Study 10.

10.2 PART 1: THE BASIC SCENARIO OF A MONOCENTRIC CITY

The basic scenario replicates the monocentric model with all basic employment (say, 100) concentrated at the CBD tract, and $\alpha = 1.0$ and $\beta = 1.0$ for the two distance friction coefficients in the gravity kernels.

Step 1. Computing network travel time matrix in the basic scenario: Create a new workflow `Case10`. Add two GeoFile Reader nodes (Nodes 1 and 2) to the Editor and input the file paths for `trtpt.zip` and `road.zip` in Node 1 and Node 2 (e.g., `E:\KNIME\CMGISV3\Chp10\data\trtpt.zip`, `E:\KNIME\CMGISV3\Chp10\data\road.zip`), respectively. Execute the nodes to read the data. We can use a Geospatial View node (Node 3) to visualize the road network.

Add a Road Network Distance Matrix node (Node 4), link GeoFile Reader (Node 1) to its first two input ports as both origin and destination points, and link GeoFile Reader (Node 2) to its third input port. In the dialog, choose `ORIG_FID` for both Origin ID column and Destination ID column, and choose `speed` for Road network speed column check `Miles per hour(mph)` for Speed unit. Execute it to calculate the OD distance list.

The resulting table contains the key columns, `Origin ID`, `Destination ID`, `Duration` and `Distance`, where the values in columns `Origin ID` and `Destination ID` are identical to the values of `ORIG_FID`.

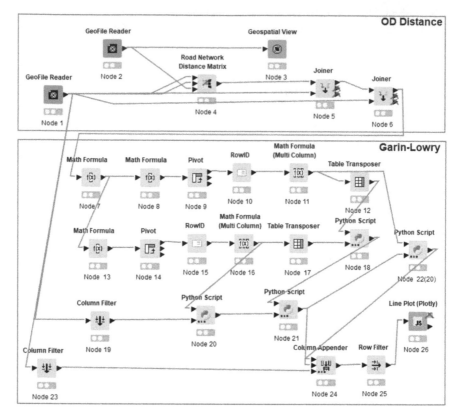

FIGURE 10.1 Workflow of Case Study 10

Step 2. Amending network travel time matrix by accounting for intrazonal travel time: We need to account for the intrazonal travel distance, e.g., approximated as ¼ of a tract's perimeter here, not only to avoid zero-distance terms in calibration of gravity model but also to capture more realistic travel impedance (Wang, 2003).

In implementation, add two Joiner nodes (Nodes 5 and 6), link Road Network Distance Matrix (Node 4) and GeoFile Reader (Node 1) to Joiner (Node 5), and then link Joiner (Node 5) and GeoFile Reader (Node 1) to Joiner (Node 6). Use Joiner (Node 5) to join column PERMITER to the OD distance list based on the common columns Origin ID and ORIG_FID, and use Joiner (Node 6) to join the column PERMITER to the updated OD distance list again based on the common columns Destination ID and ORIG_FID. The resulting table contains two columns, PERIMETER and PERIMETER(right), representing the perimeters of the origin and destination tracts in kilometers.

Add a Math Formula (Node 7) and link Joiner (Node 6) to it. Configure it to append a column `netwkTime` with the expression (`$PERIMETER$+$PERIMETER (right)$`)`*0.25*1000/804.67+$Duration$`. Note that 1 mph = 26.8224 meters/minute, and thus 30 mph = 30*26.8224 = 804.67 meter/minutes.

Step 3. Calculating gravity kernel matrix ***T***: Add a Math Formula node (Node 8) and link Math Formula (Node 7) to it. Configure it to append a column `G` with the expression `pow($netwkTime$,-1)`, which corresponds to the value of $d_{ij}^{-\alpha}$ in equation 10.2 in the main book.

Add a Pivot node (Node 9) and link Math Formula (Node 8) to it. In the dialog, (1) under Group tab, choose `Origin ID` for Group column(s), (2) under Pivots tab, choose `Destination ID` for Pivot column(s), (3) under Manual Aggregation tab, choose `G` for Column and `First` for Aggregation, (4) under Advanced settings, choose `Pivot name` for Column name and `Keep original names` for Aggregation name, and keep other default settings. Click OK and execute it to transform the OD list to OD matrix. The Pivot table port of Pivot (Node 9) contains the group column `Origin ID` and columns 1–136, derived from Destination ID.

Add a RowID node (Node 10) and link the Pivot table port of Pivot (Node 9) to it. Configure it to replace the Row ID with the column `Origin ID`, and remember to check `Remove selected column` to delete the column `Origin ID`. The result table is the matrix form of $d_{ij}^{-\alpha}$.

Add a Math Formula (Multi Column) node (Node 11) and link RowID (Node 10) to it. In the dialog, choose all columns in the Include box, input the Expression `$$CURRENT_COLUMN$$/COL_SUM($$CURRENT_COLUMN$$)*0.3`, and check Replace Selected Columns. Click OK and execute it to calculate $ed_{ij}^{-\alpha} / \sum_{l=1}^{n} d_{ij}^{-\alpha}$ in Equation 10.2 in the main book.

To get the gravity kernel **T** based on the distance matrix from destinations to origins, add a Table Transposer node (Node 12), and link Math Formula (Multi Column) (Node 11) to it. Execute it directly with the default setting to transpose it.

Step 4. Calculating gravity kernel matrix **G**: Calculating the gravity kernel matrix **G** shares the same process as calculating the gravity kernel matrix **T**. Therefore, we can reuse the nodes in the previous step.

Copy the node cluster (Nodes 8–12) and paste it as a new node cluster (Nodes 13–17). Link Math Formula (Node 7) to the copied Math Formula (Node 13) to connect the nodes. Keep the same settings in all the nodes, except for Math Formula (Multi Column) (Node 16). In the dialog of Node 16, revise the original expression as `$$CURRENT_COLUMN$$/ COL_SUM($$CURRENT_COLUMN$$)*2`. Execute the workflow to calculate the gravity kernel matrix **G**, which is $hd_{ij}^{-\beta} / \sum_{k=1}^{n} d_{ik}^{-\beta}$ in Equation 10.3 in the main book.

Step 5. Implementing matrix operation for ***I-GT*** *with Python Scripts node*: Add a Python Script node (Node 18) and configure it to have two input ports and one output port. Link the two Table Transposer nodes (Nodes 12 and 17) to it. In the dialog of

TABLE 10.1

Python Script for Calculating Matrix Operation I-GT (Node 18)

Codes	Explanation
1	import knime.scripting.io as knio import numpy as np import pandas as pd	Import packages
2	A = knio.input_tables[0].to_pandas() B = knio.input_tables[1].to_pandas()	Import data
3	T = A.values G = B.values	Table to matrices
4	N = len(A)	Get row number
5	IGT = np.identity(N)—np.dot(G,T)	Calculate **I-GT**
6	df = pd.DataFrame(IAB)	Matrix to table
7	knio.output_tables[0] = knio.Table. from_pandas(df)	Export data

Python Script node (Node 18), input the codes provided in Table 10.1 (or read from the file `I-GTmatrix.py`) to implement the matrix operation **I-GT** in Equation 10.6 in the main book.

*Step 6. Implementing matrix multiplication for **GB** with Python Scripts node*: Add a Column Filter node (Node 19) and link GeoFile Reader (Node 1) to it. Configure it to keep only one column `BEMP_CBD`, which corresponds to the value of **B** in Equation 10.5 in the basic scenario.

Copy Python Script (Node 18) as a new node Python Script (Node 20), and link Math Formula (Multi Column) (Node 16) and Column Filter (Node 19) to it. In the dialog, revise the codes according to Table 10.2 (or read from the file `Matrixdot.py`). The result is the matrix multiplication of **GB** in Equation 10.6 in the main book.

Step 7. Solving the system of linear equations (SLE) with Python Scripts node: Copy Python Script (Node 18) as a new node Python Script (Node 21), and link the two Python Script nodes (Nodes 18 and 20) to it. In the dialog, revise the code according to Table 10.3 (or read from the file `SLE.py`). Solving the SLE utilizes the functions *lu_factor* and *lu_solve* in the Python package *scipy*, and yields the population pattern in vector **P** in Equation 10.6 in the main book.

Copy Python Script (Node 20) and paste it as a new node Python Script (Node 22). Link Math Formula (Multi Column) (Node 11) and Python Script (Node 21) to it. Execute the workflow to implement the matrix multiplication **TP** to yield the service employment pattern in vector **S** in Equation 10.4 in the main book.

Step 8. Exploring population and service employment patterns in the basic scenario: Add a Column Filer node (Node 23) and link GeoFile Reader (Node 1) to it. The Column Filter is configured to keep only one column, `TRACTO_ID`.

TABLE 10.2
Python Script for Calculating Matrix Operation GB (Node 20)

	Codes	Explanation
1	import knime.scripting.io as knio import numpy as np import pandas as pd	Import packages
2	A = knio.input_tables[0].to_pandas() B = knio.input_tables[1].to_pandas()	Import data
3	G = A.values B = B.values	Table to matrices
4	GB = np.dot(G, B)	Calculate **GB**
5	df = pd.DataFrame(GB)	Matrix to table
6	knio.output_tables[0] = knio.Table.from_pandas(df)	Export data

TABLE 10.3
Python Script for Solving a System of Linear Equations (Node 21)

	Codes	Explanation
1	import knime.scripting.io as knio import numpy as np import pandas as pd from scipy.linalg import lu_factor, lu_solve	Import packages
2	A = knio.input_tables[0].to_pandas() B = knio.input_tables[1].to_pandas()	Import data
3	IGT = A.values PB = B.values	Table to matrices
4	N = len(IGT)	Get row number
5	lu, piv = lu_factor(IGT.T)	Factor matrix
6	x = lu_solve((lu, piv), PB)	Solve
5	df = pd.DataFrame(x)	Matrix to table
6	knio.output_tables[0] = knio.Table.from_pandas(df)	Export data

Add a Column Appender node (Node 24), which is configured to have three input ports. Link Python Script (Nodes 22 and 21) and Column Filer (Node 23) to Column Appender (Node 24). Execute it to combine all three tables.

Since the values for tracts on the same ring are identical, we select 10 tracts from different rings along the same direction. Use a Row Filter node (Node 25) to keep the records with TRACT0_ID values ranging from 1 to 10 by inputting 1 for lower bound, 10 for upper bound. Then, use a Line Plot (Plotly)[1] node to explore the changing trends of TRACT0_ID and other two columns, 0 and 0(#1), which represent population and service employment, respectively (Figure 10.2). Figure 10.3a-b shows the population and service employment patterns.

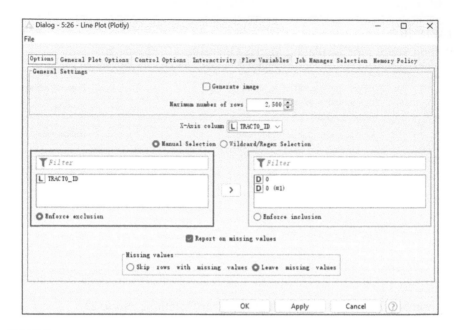

FIGURE 10.2   Interface of Line Plot (Plotly)

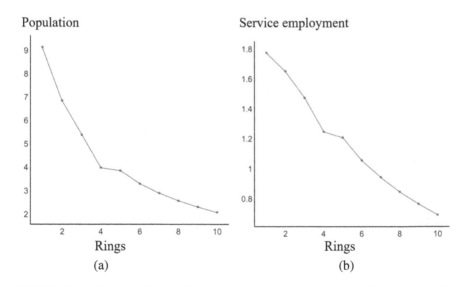

FIGURE 10.3   (a) Population and (b) Service employment cross rings in the basic scenario

10.3   PART 2: EXPLORING OTHER SCENARIOS

Step 9. Examining the impact of basic employment pattern: In the basic scenario discussed in Part 1, all basic employment is concentrated in the CBD as a monocentric city. Here we begin to examine the impact of changing the basic employment pattern to a uniform distribution. In this case, all tracts have the same amount of basic employment, i.e., 100/136 = 0.7353, which has been predefined in the field `BEMP_Unif`.

Simply reconfigure the Column Filter (Node 19) by keeping only the column `BEMP_Unif` and then re-execute the entire workflow.

Step 10. Examining the impact of travel friction coefficients: Keep all settings in the basic scenario (including the monocentric basic employment pattern) except changing the travel friction coefficients from $\alpha = 1$ and $\beta = 1$ to $\alpha = 2$ and $\beta = 2$.

To implement this change, simply reconfigure the two Math Formula nodes (Nodes 8 and 13) by revising the expression to `pow($netwkTime$,-2)`. Then re-execute the entire workflow to view the updated result table and plots.

Step 11. Examining the impact of transportation network: Here, we examine the impact of building a suburban beltway, e.g., the travel speed on the 7th ring road increases to 75 miles per hour (mph) while keeping all other roads at 30 mph. To implement this, we only need to reconfigure the Road Network Distance Matrix (Node 4), choose `speed1` for Speed column from road network and re-execute the workflow to implement this scenario.

10.4   CONCLUDING REMARKS

This chapter introduces the method for solving a system of linear equations (SLE). It is used to implement the Garin-Lowry model, commonly used by urban planners and geographers for analyzing urban land use structure. The case study on a hypothetical city illustrates how the Garin-Lowry model reveals the interaction between employment and residential (population) patterns and how the basic employment plays a different role from the nonbasic employment. Various scenarios are simulated to examine the impacts of basic employment patterns, distance friction coefficients and road network improvements (e.g., construction of a suburban beltway).

The workflow utilizes several nodes, including Math Formula and Rule Engine, to define the speed parameter in calibrating an OD matrix via a street network. Pivot and Math Formula (Multi Column) nodes are used to define the gravity kernels based on the OD matrix.

All matrix manipulations and equation solutions are implemented using Python Scripts, which rely on simple codes based on functions such as *identity()* and *np.dot()* from the Numpy package, and *lu_factor()* and *lu_solve()* from the Scipy package.

NOTE

1   The Line Plot (Plotly) node automatically sorts the x-axis values, and has a better performance than Line Plot (JavaScript). User might need to install the JavaScript Views (Labs) extension to use this node.

11 Linear and Quadratic Programming and Applications in Examining Wasteful Commuting and Allocating Healthcare Providers

This chapter introduces two popular methods in optimization, *linear programming* (LP) and *quadratic programming* (QP). QP is perhaps the simplest form of *non-linear programming* (NLP). LP and QP seek to maximize or minimize an objective function subject to a set of constraints. LP has both the objective and the constraints in linear functions. QP has a quadratic objective function, but its constraints remain linear. This chapter uses case studies to illustrate their applications in spatial planning and location-allocation analysis.

Case Study 11A examines the issue of *wasteful commuting* in Columbus, Ohio to illustrate the formulation of LP and its solution. Case Study 11B uses various *location-allocation problems* to plan healthcare providers in a rural county in China. In addition to data preparation, one part illustrates some classic location-allocation models such as the *p*-median problem, the maximum covering location problem (MCLP), and the minimax problem that belong to the family of *integer linear programming (ILP)* problems. The other part introduces a new location-allocation problem termed "*Maximal Accessibility Equality Problem (MAEP)*" and demonstrates how the problem is solved by QP. The MAEP seeks to minimize inequality in accessibility of facilities across geographic areas, and by extension, across population groups, and has great potential in applications in both private and public sectors.

11.1 CASE STUDY 11A: MEASURING WASTEFUL COMMUTING IN COLUMBUS, OHIO

Recall the definition for wasteful commuting. Given the number of resident workers P_i at i ($i = 1, 2, \ldots, n$) and the number of jobs at E_j at j ($j = 1, 2, \ldots, m$) in a city, the minimum commute is the solution to the following linear programming problem:

DOI: 10.1201/9781003304357-11

Minimize: $\sum_{i=1}^{n}\sum_{j=1}^{m}C_{ij}x_{ij}$

Subject to: $\sum_{j=1}^{m}x_{ij} \leq P_i$ for all $i\ (=1, 2, \ldots, n)$

$\sum_{i=1}^{n}x_{ij} \leq E_j$ for all $j\ (=1, 2, \ldots, m)$

$x_{ij} > 0$ for all $i\ (=1, 2, \ldots, n)$ and all $j\ (=1, 2, \ldots, m)$

where C_{ij} is the commute distance (time) from residential location i to job site j, and x_{ij} is the number of commuters on that route to be solved.

The objective function is the total amount of commute in the city. The first constraint defines that the total commuters from each residential location to various job locations cannot exceed the number of resident workers there. The second constraint defines that the total commuters from various residential locations to each job site cannot exceed the number of jobs there. In the urbanized areas of most U.S. metropolitan areas, it is most likely that the total number of jobs exceeds the total number of resident workers, i.e., $\sum_{i=1}^{n}P_i \leq \sum_{j=1}^{m}E_j$.

Data used in this case study is based on a study reported in Wang (2001). The data folder `Columbus` includes:

1. a zipped area feature class `urbtaz.zip` and its corresponding point feature class `urbtazpt.zip` with 991 TAZs (traffic analysis zones and their centroids, respectively)
2. a zipped feature dataset `roads.zip` containing a single feature class roads for the road network, and
3. R file `WasteCommute.r` for measuring waste commute.

The fields WORK and EMP in `urbtazpt` define the numbers of resident workers P_i $(i = 1, 2, \ldots, n)$ and employment E_j $(j = 1, 2, \ldots, m)$ in each TAZ, respectively.

Figure 11.1 illustrates the workflow of Case Study 11A.

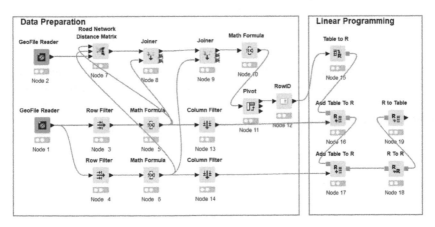

FIGURE 11.1 Workflow of Case Study 11A

Step 1. Extracting locations of non-zero employment and non-zero resident workers: For improving computational efficiency, we need to restrict the origins and destinations to those TAZs with nonzero employment and nonzero resident worker counts, respectively, when computing the OD travel time C_{ij}.

Create a new workflow Case11A. Add two GeoFile Reader nodes (Nodes 1 and 2) to the Editor and input the file paths for census tract points (urbtazpt.zip) and road network (roads.zip) in Node 1 and Node 2, respectively (e.g., E:\KNIME\ CMGISV3\Chp11\data\urbtazpt.zip, E:\KNIME\CMGISV3\Chp11\data\ roads.zip). Execute the nodes to read the data.

Add two Row Filter nodes (Nodes 3 and 4) and link GeoFile Reader (Node 1) to them. In the dialog of Row Filter (Node 3), check Include rows by attribute value, choose WORK for Column to test, check use range checking and input 1 for lower bound. Click OK and execute it to keep the TAZs with WORK ≥ 1 (812 resident worker locations retained). Configure Row Filter (Node 4) to keep the TAZs with EMP ≥ 1 (931 employment locations retained). They will be used as the origins and destinations to generate the OD time matrix.

Step 2. Computing the OD time matrix between resident workers and employment locations: Add two Math Formula nodes (Nodes 5 and 6), and link Row Filter (Node 3) to Math Formula (Node 5) and Row Filter (Node 4) to Math Formula (Node 6). Configure them to append a new column OID (DID) with the expression $$ROWINDEX$$+1, which assigns new IDs to be used in network analysis.

Add a Road Network Distance Matrix node (Node 7). Link the two Math Formula nodes (Nodes 5 and 6) to the first two input ports of Node 7 as both origin and destination points, and link GeoFile Reader (Node 2) to the third input port of Node 7. In the dialog, choose OID for Origin ID column and DID for Destination ID column, choose Speed for Road network speed column and check Miles per hour(mph) for Speed unit. Execute it to calculate the OD distance list. This study uses a simple approach for measuring travel time by assuming a uniform speed on the same level of roads.

The resulting table contains six columns, such as Origin ID, Destination ID, Duration and Distance, where the values of Origin ID and Destination ID are identical to those of OID and DID, respectively.

Similar to step 2 in Section 10.3 of Chapter 10, intra-zonal time here is approximated as ¼ of a TAZ's perimeter divided by a constant speed 670.56 meters/ minute (i.e., 25 mph) in this study. The total travel time between two TAZs is composed of the network time from step 1 and the intra-zonal time at both origin and destination TAZs.

Add two Joiner nodes (Nodes 8 and 9), link Street Network Matrix node (Node 7) and Math Formula (Node 5) to Joiner (Node 8), then link Joiner (Node 8) and Math Formula (Node 6) to Joiner (Node 9). Configure the two Joiner nodes to attach column length based on the columns Origin ID (OID) and Destination ID (DID), where the duplicated column length is automatically renamed as length(right).

Add a Math Formula node (Node 10) and link Joiner (Node 9) to it. Configure it to update the values in the column `Duration` with the expression `$Duration$+0.25*($Length$+$Length(right)$)/670.56`. This amends the network travel time by adding intra-zonal times.

Step 3. Converting OD list to OD matrix: Add a Pivot node (Node 11) and link Math Formula (Node 10) to it. In the dialog, (1) under Group tab, choose `Origin ID` for Group column(s), (2) under Pivots tab, choose `Destination ID` for Pivot column(s), (3) under Manual Aggregation tab, choose `Duration` for Column and `First` for Aggregation, (4) under Advanced settings, choose `Pivot name` for Column name and `Keep original names` for Aggregation name, and keep other default settings. Click OK and execute it to transform the OD list to OD matrix. The Pivot table port of Node 11 contains the group column `Origin ID` and the columns 1–931, derived from `Destination ID`.

Add a RowID node (Node 12) and link the Pivot table port of Node 11 to it. Configure it to replace the Row ID with the column `Origin ID`, and remember to check `Remove selected column` to delete the column `Origin ID`. The result table is the travel time matrix for C_{ij}.

To prepare the vector for resident workers P_i at i ($i = 1, 2, \ldots, n$) and the vector for jobs E_j at j ($j = 1, 2, \ldots, m$), add two Column Filter nodes (Nodes 13 and 14), and link Math Formula (Node 5) to Column Filter (Node 13) and Math Formula (Node 6) to Column Filter (Node 14). Configure both Column Filter nodes to keep only one column each: `WORK` for Column Filter (Node 13) and `EMP` for Column Filter (Node 14).

A simple example in Figure 11.2 illustrates the LP problem for measuring wasteful commuting. A vector of 4 resident worker locations (P_i) intersects with a vector of 5 employment locations (E_j) to yield a 4×5 OD travel time matrix C_{ij}. The variable to be solved is elements t_{ij} as number of commuters from i to j. The *lpSolve* package in R is used to solve the problem. The solution is shown as red numbers.

Step 4. Reading travel time, resident workers and employment data in matrix (vector) to R: Add one Table to R node (Node 15), and two Add Table To R nodes (Nodes 16 and 17). Link RowID (Node 12) to Table to R (Node 15); link the R output

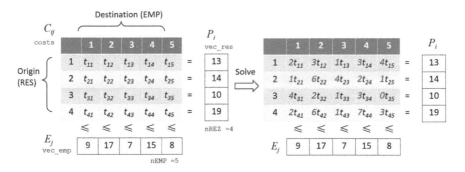

FIGURE 11.2 An illustrative example for the wasteful commuting problem

of Table to R (Node 15) and Column Filter (Node 13) to the corresponding ports of Add Table To R (Node 16), and connect the R output of Add Table to R (Node 16) and Column Filter (Node 14) to Add Table To R (Node 17). Input the three lines in Nodes 15–17, respectively, to read all data and transform them into matrix in the R workspace:

```
costs<-as.matrix(knime.in)
vec_res<-t(as.matrix(knime.in))
vec_emp<-t(as.matrix(knime.in))
```

wherein, `knime.in` is the input port data, `as.matrix` transforms the data to matrix, and `t` transposes data.

Step 5. Solving LP problem in R: Add an R To R node (Node 18) and link Add Table To R (Node 17) to it. Input the R codes in the file `WasteCommute.r` provided in Table 11.1 in the node, where the comments explain the codes.

Add an R to Table node (Node 19) and link it to R To R (Node 18). Input the two-line codes in Node 19:

```
knime.out <- data.frame(OD=as.vector(lpresult$solution))
print(paste('wasteful commuting is ',wastc))
```

It exports the data via the output port and prints the result in "View: R Std Output". The results show that the average commute time is 20.12 minutes, the average optimum commute time is 7.679 minutes, and thus wasteful commuting is 61.833%. Further examination of the optimal commuting patterns in the output table of R to Table (Node 19) reveals that many trips have the same origin and destination

---

TABLE 11.1

Codes for Solving LP Problem in R To R

Code	Explanation
require(lpSolve) require(dplyr)	Load package
nREZ < -length(vec_res) nEMP <- length(vec_emp)	The numbers of origin and destination points
row.signs <- rep("=",nREZ) row.rhs <- vec_res	$\sum_{j=1}^{m} x_{ij} \leq P_i$ for all $i (= 1, 2, \ldots, n)$
col.signs <- rep("<=",nEMP) col.rhs <- vec_emp	$\sum_{i=1}^{n} x_{ij} \leq E_j$ for all $j (= 1, 2, \ldots, m)$
lpresult<-lp.transport(costs,"min", row.signs,row.rhs,col.signs,col.rhs)	Minimize: $\sum_{i=1}^{n} \sum_{j=1}^{m} C_{ij} x_{ij}$
optimc<-lpresult$objval/sum(vec_res)	Average optimum commute time
wastc<- (mean(costs)-optimc)/mean(costs)	wasteful commuting ratio

TAZs. Therefore, accounting for intrazonal commute time is significant in reducing wasteful commuting. A significant portion of so-called wasteful commuting may be attributable to inaccurate travel time estimation.

11.2 CASE STUDY 11B: LOCATION-ALLOCATION ANALYSIS OF HOSPITALS IN RURAL CHINA

This case study is developed from the work reported in Luo et al. (2017). The planning problem is where to build three new hospitals and how large for each. The study area is a rural county Xiantao in Hubei Province of China. A sequential decision-making approach, termed *"two-step optimization for spatial accessibility improvement (2SO4SAI)"*, is conceptualized to solve the problem.

The case study is implemented in three parts. Part 1 prepares the data. Part 2 finds the best locations to site the new hospitals by emphasizing accessibility as proximity to the nearest hospitals. Part 3 adjusts their capacities for minimal inequality in accessibility measured by the 2SFCA method. Part 2 strikes a balance among the solutions by three classic location-allocation models (*p*-median, MCLP, and minimax problems), and Part 3 solves a QP problem.

The data folder Xiantao includes:

1. Supply point feature class Hosp41.zip includes 41 existing hospitals with a field CHCI (comprehensive hospital capacity index) representing their capacities, and HospAll.zip adds 3 newly sited hospitals to Hosp41 as a result from Part 2.
2. Demand point feature class Village.zip contains 647 villages with a field Popu indicating their population sizes.
3. Feature dataset Road.zip represents the major road network.
4. Two base layers include Xiantao.zip for the study area boundary and Township.zip for the township administrative boundaries (each township includes multiple villages).
5. Intermediate results include (1) a distance table ODhosp41.csv from 647 villages to extant 41 hospitals, and (2) another distance table ODvillage. csv from 647 villages to four candidate hospital sites.

Figure 11.3 shows the workflow for Case Study 11B.

11.2.1 PART 1: DATA PREPARATION

Step 1. Calculating OD matrix between villages and existing hospitals: Create a new workflow Case11B1. Add three GeoFile Reader nodes (Nodes 1–3) to the Editor and input the file paths for the village data Village.zip, hospital data Hosp41.zip and road network Road.zip, respectively (e.g., E:\KNIME\ CMGISV3\Chp11\data\Xiantao\Village.zip). Execute the nodes to read the data.

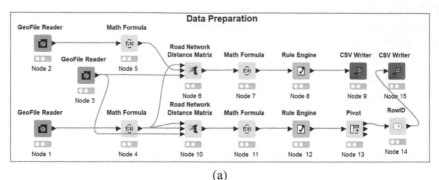

(a)

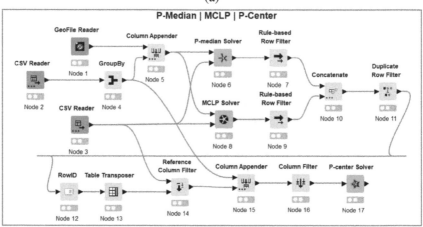

(b)

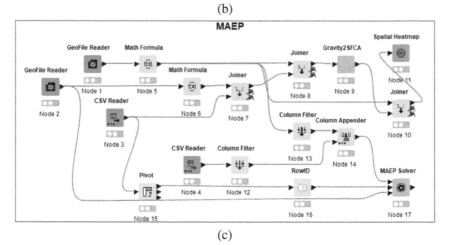

(c)

FIGURE 11.3 Workflow of Case Study 11B: (a) Part 1, (b) Part 2, and (c) Part 3

Add two Math Formula nodes (Nodes 4 and 5), link GeoFile Reader (Node 1) to Math Formula (Node 4), and link GeoFile Reader (Node 2) to Math Formula (Node 5). Configure them to append a new column OID (DID) with the expression $$ROWINDEX$$+1, which assigns new IDs for network analysis.

Add a Road Network Distance Matrix node (Node 6), link the two Math Formula nodes (Nodes 4 and 5) to its first two input ports as origin and destination points, respectively, and link GeoFile Reader (Node 3) to its third input port. In the dialog, choose OID for Origin ID column and DID for Destination ID column, choose Speed for Road network speed column from road network, and check Kilometers per hour(km/h) for Speed unit. Execute it to calculate the OD distance list ($647 \times 41 = 26{,}727$ records).

Add a Math Formula node (Node 7) and link Road Network Distance Matrix (Node 6) to it. Configure it to update the values in column Distance with the expression $Distance$+$Origin snap distance$+$Destination snap distance$. This amends the network travel time by adding the snap distances at the two ends.

To avoid zero-distance values in the subsequent distance decay calculation, we need to set a minimum distance of 300 meters between hospitals and villages. Add a Rule Engine node (Node 8) and link it to the Math Formula node (Node 7). The nodes can be configured to replace the values of Distance using the following expression:

```
$Distance$ > 300=> $Distance$
$Distance$ <= 300=> 300
```

Add a CSV Writer node (Node 9) and configure it to save the distance matrix file as a CVS file such as ODhosp41.csv (e.g., E:\KNIME\CMGISV3\Chp11\data\ODhosp41.csv).

Step 2. Calculating OD matrix between villages: This step essentially replicates Step 1 except that the distances are from villages to villages themselves. Since the new hospital sites come from any of the villages,[1] the derived distance matrix defines distances from demand locations to candidate hospital sites.

Copy the three nodes (Nodes 6–8) along with their links and paste them as new nodes (Nodes 10–12). Link Math Formula (Node 4) to the first two input ports of Road Network Distance Matrix (Node 10) as both the origin and destination points, and connect GeoFile Reader (Node 3) to the third input port for the road network. Then, run the three nodes with the same settings as in Step 1 to generate an OD distance matrix with a default minimum distance of 300 meters ($647 \times 647 = 418{,}609$ records).

Add a Pivot node and a RowID node, and link Rule Engine (Node 12) $\rightarrow$ Pivot (Node 13) $\rightarrow$ RowID (Node 14) to transform the OD distance list as OD matrix. Use Origin ID as row ID and delete the column Origin ID. Refer to Step 3 in Section 11.1 for detailed configurations.

Add a CSV Writer (Node 15) to export the result table and save it as ODVillage.csv (e.g., E:\KNIME\CMGISV3\Chp11\data\ODVillage.csv).

11.2.2 PART 2: LOCATION OPTIMIZATION FOR SITE SELECTION

For all three location-allocation models for site selection, existing facilities are hospitals defined in spatial data `Hosp41` with their capacities in the field `CHCI`, and demands are villages in spatial data Village with their demand amount in the field `Popu`. As stated previously, candidate facilities come from the village locations. The objective of location optimization is to select three village locations out of 647 villages as new hospital locations while taking into account of the presence of 41 existing hospitals.

Figure 11.4 uses an illustrative example to show what data sets are prepared for the optimization problems. For example, as shown in Figure 11.4a, one needs to select two from a pool of three *candidate facilities* (A, B, and C) to serve four *demand points* (1, 2, 3 and 4), while two *required facilities* (D and E) already exist in place. The solution may end up with *chosen facilities* A and B.

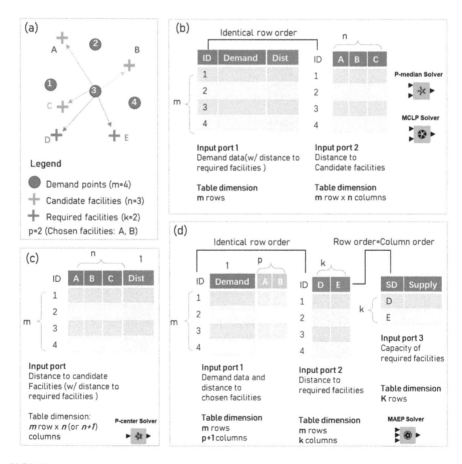

FIGURE 11.4 Illustrations for Case Study 11B input data: (a) data components, (b) p-median or MCLP Solver, (c) p-center Solver, and (d) MAEP Solver

Step 3. Preparing data for P-Median Solver and MCLP Solver: Figure 11.4b illustrates that both the P-median Solver and MCLP Solver nodes require two types of input data: (1) demand feature includes a population column and an optional column for recording the distance to the nearest required facility; and (2) distance matrix specifies the distances between demand points and candidate facilities.

Create a new workflow Case11B2. Copy the GeoFile Reader nodes (Node 1) in Case11B1 to the Editor to read village data Village.zip. This prepares the demand (population) dataset.

Add two CSV Reader nodes (Node 2 and Node 3) and configure them to read the OD distance matrix file ODhosp41.csv (between villages) in Node 3 (647 × 647 = 418,609 columns) and another OD distance list file ODVillage.csv (from villages to existing or required hospitals) in Node 2 (647 × 41 = 26527 rows). The data read by Node 3 contains columns 1–647 for the distances between all villages, which will be directly fed into input port 2 of the two solver nodes. The data read by Node 2 needs to be processed in order to prepare a distance column for all required (extant) facilities as dummy facilities.

Add a GroupBy node (Node 4) and link CSV Reader (Node 2) to it to summarize distances to the nearest hospitals by village IDs. In the dialog, (1) under the Groups tab, choose Origin ID for Group column(s) and Keep original name(s) for Column naming, (2) under the Manual Aggregation tab, choose Distance for Column and Minimum for Aggregation. Click OK and execute it. The column Distance in the resulting table represents the distance to the nearest hospital for each village. This creates a dummy facility dataset that represents all 41 existing hospitals.

Add a Column Appender node (Node 5). Link GeoFile Reader (Node 1) and GroupBy (Node 4) to the two input ports of Column Appender (Node 5). Execute it to combine the two datasets, which will be fed into the first input port of the two Solver nodes. If the two input data sets share the same order but have different row IDs, check Generate new row keys in the dialog before executing the node to ensure that the data are properly combined. The resulting table contains two columns: Popu for population and Distance for the dummy facility representing the nearest hospital.

Step 4. Applying P-Median Solver and MCLP Solver: Under Location Analysis, drag a P-Median Solver node (Node 6) to the Editor. Link Column Appender (Node 5) and CSV Reader (Node 3) to it. In the dialog of P-Median Solver (Figure 11.5a), input 3 for Optimum number of facilities (p), Popu for Demand column, Distance for Cost column for required facilities, click OK and execute it. The resulting table contains (1) column FacilityID identical to the column names 1–647 of the first input data (dummy facility marked as required), and (2) column Chosen with values 1 identifying chosen facilities.

Add a Rule-based Row Filter node (Node 7) to export the chosen facilities and filter out the dummy facilities by its index with the following expression $Chosen$ =1 AND $$ROWINDEX$$ < $$ROWCOUNT$$=> TRUE.

Add a MCLP Solver node (Node 8), and link Column Appender (Node 5) and CSV Reader (Node 3) to it. In the dialog box (Figure 11.5b), configure it with the same parameters as the P-median Solver (Node 6), and input 20000 (20 km) for

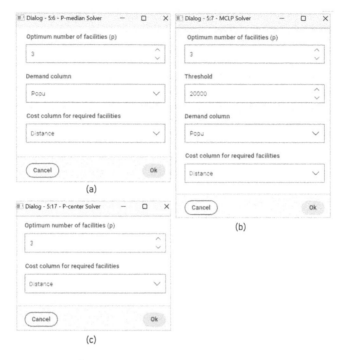

FIGURE 11.5 Interfaces[2] of (a) P-median Solver, (b) MCLP Solver, and (c) P-center Solver

the additional parameter Threshold. The result table has the same columns as the P-median Solver node.

Copy Rule-based Row Filter (Node 7) and paste it as a new node (Node 9). Link MCLP Solver (Node 8) to Node 9 and execute it to keep the chosen facility IDs.

Next, add a Concatenate node (Node 10) and link the two Rule-based Row Filter nodes (Nodes 7 and 9) to it. Execute it to combine the chosen locations to serve as candidate facilities for the P-center model.

Due to the possibility of overlap between locations chosen by the P-median and MCLP solvers, it is necessary to use a Duplicate Row Filter (Node 11) to remove any duplicates. The default settings for the filter can be used to achieve this. In this scenario with the current settings, it yields 4 unique chosen facilities.

Step 5. Applying P-center Solver: As shown in Figure 11.5c, to apply the P-center Solver, we need to build an OD matrix between the villages and all facilities (including candidate facilities and dummy facilities). As stated in the main book, we limit the candidate choices for the P center solver to the four facilities yielded from step 4 to demonstrate the process and save computational time.

Add a RowID node (Node 12) and link Duplicate Row Filter (Node 11) to it. In the dialog shown in Figure 11.6, configure it to choose `FacilityID` as Row IDs and remove the column `FacilityID`. Then use a Table Transposer node (Node 13) to transpose the data, which automatically sets the values in `FacilityID` as the column names.

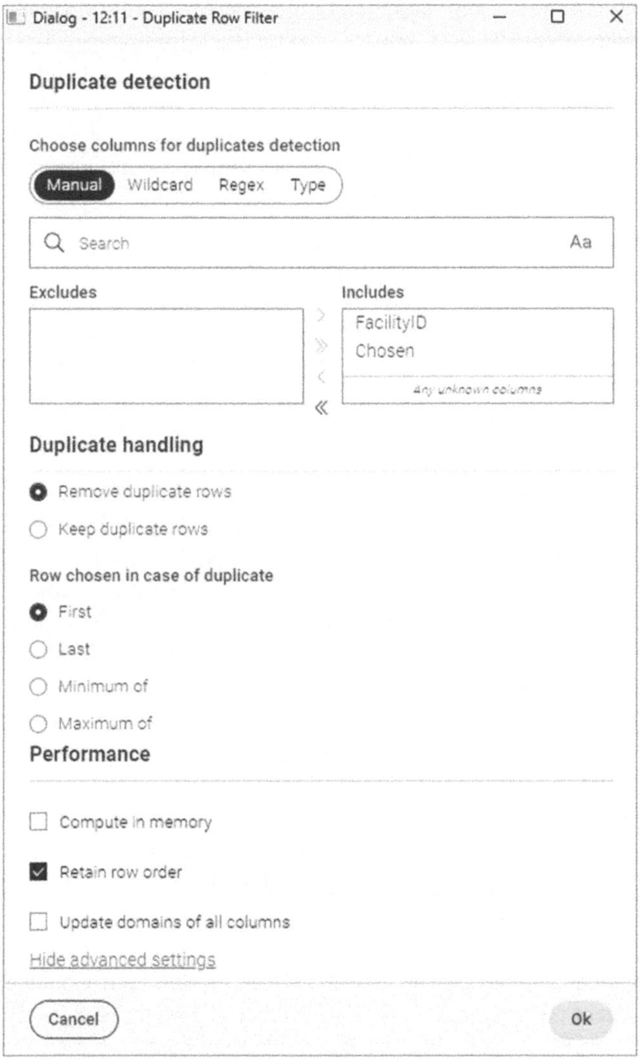

FIGURE 11.6   Interfaces of Duplicate Row Filter

Add a Reference Column Filter node (Node 14), link CSV Reader (Node 3) and Table Transposer (Node 13) to it. In the dialog, check `Include` under Include or exclude columns from reference table. Execute it to get the corresponding OD matrix between villages and the chosen facilities from the full OD matrix.

Similar to Step 4, add a Column Appender node (Node 15) and link GroupBy (Node 4) and Reference Column Filter (Node 14) to it. Execute it to combine the two datasets.

Add a Column Filter node (Node 16) and link Column Appender (Node 15) to it. Configure it to exclude the irrelevant column `Origin  ID` from the data of GroupBy (Node 4) and only keep the distances for chosen facilities and the required dummy facilities.

Next, add a P-center Solver node (Node 17) and link Column Filter (Node 16) to it. In the dialog (Figure 11.5c), choose 3 for Optimum number of facilities (p), and choose Distance for Cost column for required facilities. Execute it to apply the P-center model.

It yields three chosen sites (IDs: 74, 401, 435), which will be the newly sited hospitals for the MAEP to build upon and look for optimal capacities to assign in Part 3.

11.2.3 Part 3: Implementing the MAEP to Derive Capacities for Sited New Hospitals

Step 6. Assessing disparity in gravity-based 2SFCA accessibility scores: The MAEP adjusts the allocation of total hospital capacity among all hospitals or limited only to those selected new facilities in order to maximize the equality of accessibility across demands. It helps to assess the disparity of accessibility prior to the adjustment.

Create a new workflow Case11B3. Copy GeoFile Reader nodes (Nodes 1 and 2) in Case11B1 to read the geospatial data Village.zip and Hosp41.zip. Copy CSV Reader nodes (Nodes 2 and 3) in Case11B2 as Nodes 3 and 4 to read the OD distance matrix ODhosp41.csv and ODVillage.csv. Copy the two Math Formula nodes (Nodes 4 and 5) in Case11B1 as Nodes 5 and 6, link GeoFile Reader (Node 1) to Math Formula (Node 5), link GeoFile Reader (Node 2) to Math Formula (Node 6) to attach the ID columns, OID and DID, respectively.

For illustration, we use the popular power function for distance decay, $f(d_{ij}) = d_{ij}^{-\beta}$, and $\beta = 1$ in the gravity-based 2SFCA accessibility measure in Chapter 5.

Add two Joiner nodes (Nodes 7 and 8), link Math Formula (Node 6) and CSV Reader (Node 3) to Joiner Node 7, then link Math Formula (Node 5) and Joiner (Node 7) to Joiner Node 8. Configure the two Joiner nodes to attach columns Popu and CHCI based on the common columns Origin ID (OID) and Destination ID (DID).

Copy the nodes for the gravity based 2SFCA model from Chapter 5 and make it as a component Gravity2SFCA (Node 9). Link Joiner (Node 8) to the component Gravity2SFCA (Node 9). Configure the node based on the demand ID filed Origin ID and its value field Popu, supply ID field Destination ID and its value field CHCI, and the travel cost field Distance, and set −1 for power coefficient. Execute the component to calculate the accessibility scores. The result is in column docpopR.

Add a Joiner node (Node 10) and link Math Formula (Node 5) and the component Gravity2SFCA (Node 9) to it. Configure it to join column docpopR to the village location data based on the join columns OID and Origin ID.

Use the Spatial Heatmap node (Node 11) to apply spatial smoothing on the accessibility score and visualize the disparity across the study area (Figure 11.7). Refer to step 2 of subsection 3.3.1.

Step 7. Preparing data for MAEP Solver: The MAEP is formulated to allocate health care resources among the three selected candidate hospitals in order to

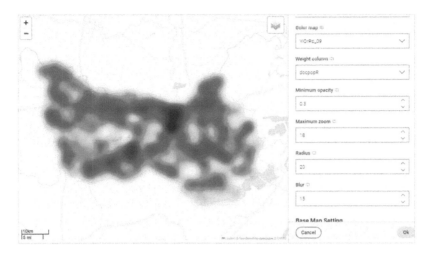

FIGURE 11.7 Gravity-based 2SFCA accessibility score heatmap

maximize accessibility equality (i.e., minimize accessibility inequality). Required data inputs include population (demand), OD matrix and hospital capacity (supply).

As depicted in Figure 11.4d, there are three input tables required for the MAEP Solver: (1) demand data and distances to new facilities, with a dimension of 647 rows × 4 columns (one column for demand population and three columns for distances from demand points to the 3 candidate facilities); (2) distances from demand locations to required (existing) facilities, with a dimension of 647 rows × 41 columns; and (3) capacity of required facilities, with a dimension of 41 rows.

Add a Column Filter node (Node 12), and link CSV Reader (Node 4) to it. In the dialog, check `Wildcard/Regex Selection`, and input 74|401|435 for the Pattern, check `Regular expression` to choose the three village IDs (i.e., newly sited three hospitals) in the Match(Include) box. Execute it to keep the three columns. This prepares a distance table between villages and the three new hospitals.

Add another Column Filter node (Node 13), and link Math Formula (Node 5) to it. Configure it to only keep the column `Popu`. Add a Column Appender node (Node 14) and link the two Column Filter nodes (Nodes 12 and 13) to it. Execute it to complete the preparation for the data needed for the first input port of MAEP node, with four columns, `Popu`, 74, 401, and 435.

Add a Pivot node (Node 15) and a RowID node (Node 16). Link CSV Reader (Node 3) → Pivot (Node 15) → RowID (Node 16) to transform the OD distance list as OD matrix, using `Origin ID` as row ID and deleting column `Origin ID`. Refer to Step 3 in Section 11.1 for detailed configurations. This prepares a table containing distances between villages and 41 existing hospitals represented by columns 1–41, which will be the data for the second input port of MAEP node.

We can now proceed with the implementation of the MAEP model, as the hospital data needed for the third input port of MAEP node is provided by the capacity column `CHCI` in GeoFile Reader (Node 2).

Step 8. Applying MAEP Solver for allocating resources: Add a MAEP Solver node (Node 17), and link Column Appender (Node 14), RowID (Node 16), and GeoFile Reader (Node 2) to it. In the dialog shown in Figure 11.8, input 7200 for New capacity, choose Popu for Demand column, CHCI for Existing facility capacity column, check Power for Distance decay model, and input 1 for Distance decay parameters. Click OK and execute it.

This node uses a power function (Gravity model) as the distance decay function with $\beta = -1$. Other distance decay functions such as exponential or 2SFCA (its associated distance threshold defines "Distance decay parameter") are also available to choose. Future refinement on the model can expand the list of choices.

The result table has two columns for two different scenarios: All represents the optimal distribution of the entire hospital capacities across all hospitals, and Fixed represents the optimal distribution of only the total new capacities among the three new hospitals while keeping the capacities of the existing hospitals intact. The last three rows of the table indicate the capacities of the three new hospitals in the two scenarios.

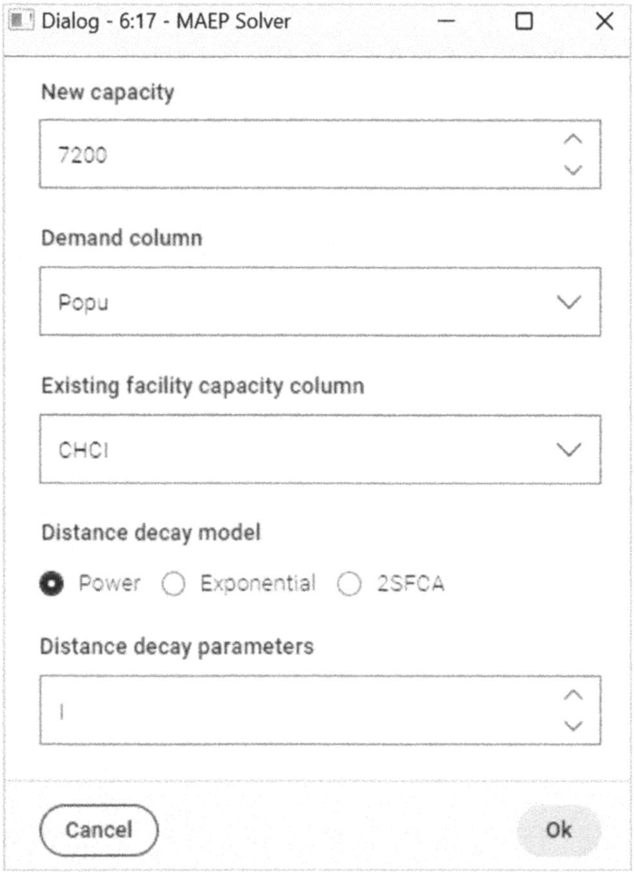

FIGURE 11.8   Interface for MAEP Solver

One may proceed to replicate Step 6 to update the accessibility measure based on the optimal allocation of resources and assess whether the disparity of accessibility is narrowed and by how much.

11.3 CONCLUDING REMARKS

This chapter covers optimization methods that include linear programming (LP), integer linear programming (ILP), and nonlinear programming (NLP) such as quadratic programming (QP). All are widely used in operational research, engineering, socioeconomic planning, and location analysis. Case Study 11A uses an application in measuring wasteful commuting to illustrate the formulation of LP. Part 2 of Case Study 11B uses classic location-allocation models such as p-median, maximum covering location problem (MCLP) and minimax to illustrate the application of ILP in site selection. Part 3 of Case Study 11B introduces a planning problem MAEP to illustrate the application of QP in allocating resources for maximal equal accessibility.

These studies are implemented by customized tools with the support of R nodes and the *lpSolve* package. Several Geospatial Analytics Extension nodes are used to solve the traditional P-media, MCLP, and P-center (Minimax) models. A new node for the Maximal Accessibility Equality Problem (MAEP) is developed to solve the quadratic programming (QP) problem.

NOTES

1 Villages are numerous and widely distributed across the study area, so it is reasonable to assume that new hospitals will be sited at (or very close to) any of the villages.
2 As the location analysis nodes are continually updated, the user interface may exhibit slight differences.

12 Monte Carlo Method and Applications in Urban Population and Traffic Simulations

Monte Carlo simulation provides a powerful computational framework for spatial analysis and has become increasingly popular with rising computing power. Some applications include data disaggregation, designing a statistical significance test and modeling individual behaviors. This chapter demonstrates the value of Monte Carlo technique in spatial analysis.

One case study uses the Monte Carlo method to simulate individual resident locations by using the census and land use inventory data, and then aggregates population back to area units of any scale in any shape. Another case study demonstrates the value of applying the Monte Carlo technique in simulating urban traffic flows. The former generates individual points, and the latter derives linkages between points. Both can benefit analysis that is prone to the scale and zonal effects. In Case Study 12A, the technique is applied to examine the role of such effects on urban population density functions. In Case Study 12B, one may use the technique of simulating individual trips to improve accuracy in travel distance estimate (e.g., reducing uncertainty in measuring wasteful commuting based on area units).

12.1 CASE STUDY 12A: DERIVING URBAN POPULATION DENSITY FUNCTIONS IN UNIFORM AREA UNIT IN CHICAGO BY MONTE CARLO SIMULATION

This case study is based on the work reported in Wang, Liu & Xu (2019). The study area is the seven-county Chicago CMSA. Main data sources include the 2010 Census data and the 2010 Land Use Inventory data. We limit the simulation of population to the residential land use category so that simulated residents can better resemble the actual settlement pattern.

As stated in section 6.3 of Chapter 6 in the main book, it is desirable to have analysis areas of identical or similar area size in fitting urban density functions. Wang, Liu & Xu (2019) designed six area units with three distinctive shapes (square, triangle, and hexagon) and two scales to capture both zonal and scale effects. This case study uses only one shape (hexagon) in one size (1 km²) for illustration. A small number of areas on the edge of the study area are truncated and thus smaller.

DOI: 10.1201/9781003304357-12

The sub-folder `ZoneEffect` under the data folder `Chicago` includes:

1) feature `residential.zip` represents residential land use,
2) feature `block.zip` represents census blocks with field `POP100` for population in 2010,
3) features `tract.zip` and `blockgroup.zip` represent census tracts and block groups, respectively, and
4) feature `citycenter.zip` is the city center.

Figure 12.1 shows the workflow of Case Study 12A.

Step 1. Allocating population to residential portion of census blocks: In KNIME AP, create a new workflow Case12A. Add two GeoFile Reader nodes to the Editor and input the file path of census blocks in Node 1 (e.g., E:\KNIME\CMGISV3\Chp04\ data\block.zip) and residential areas in Node 2 (e.g., E:\KNIME\CMGISV3\ Chp12\data\residential.zip). Execute them to read the data.

Add an Overlay node (Node 3) and link the two GeoFile Reader nodes (Nodes 1 and 2) to it. In the dialog, choose `Intersection` for the Overlay mode, and check `Return`

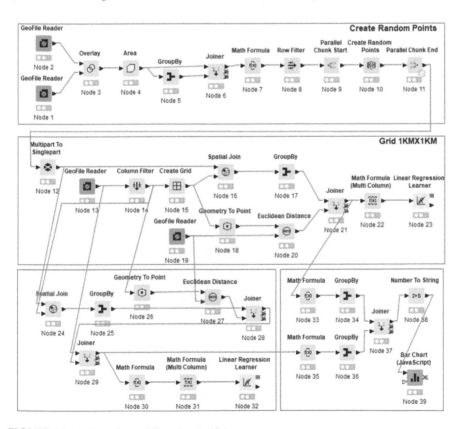

FIGURE 12.1 Workflow of Case Study 12A

only geometries of the same geometry type. The result table only contains residential portions of the census blocks, referred to as "residential blocks".

Add an Area node (Node 4) and link Overlay (Node 3) to it. Execute it to calculate the area sizes of residential blocks.

Use a GroupBy node (Node 5) to summarize the total areas by the IDs of census blocks. In the dialog, (1) under the Groups tab, choose BLOCKID10 for Group column(s) and Aggregation method(column name) for Column naming, (2) under Manual Aggregation tab, choose area for Column and Sum for Aggregation. Click OK and execute it. In the result table, the field Sum(area) represents the total areas for all units sharing the same Census Block ID.

Add a Joiner node (Node 6) and link Area (Node 4) and GroupBy (Node 5) to it. In the dialog, (1) under Joiner Settings tab, choose BLOCKID10 for both Top Input ('left' table) and Bottom Input ('right' table); (2) under Column Selection, keep all columns in the Include box for Top Input (left table) and choose Sum(area) for Bottom Input (right table), keep other default settings, and execute it. This step attaches the aggregated block areas to the intersected units.

Add a Math Formula node (Node 7) and link Joiner (Node 6) to it. Configure Math Formula (Node 7) to append an integer column Popu with the expression $area$/$Sum(area)$*$POP10$. This distributes the population in each census block to residential block(s) proportionally to their area sizes.

Finally, to ensure that all residential blocks have a non-zero population, use a Row Filter node (Node 8) to filter the records. In the dialog box, check Include row by attribute value, choose Popu for Column to test, check use range checking and input 1 for lower bound.

Step 2. Simulating random points to represent population in residential blocks: Users can utilize the Parallel Execution extension to save computational time when generating a large number of random points. Install the extension by going to File > Install KNIME Extension. Refer to Chapter 1 for more detail.

Under KNIME Labs > Parallel Execution, drag the Parallel Chunk Start (Node 9) to the Editor and link Row Filter (Node 8) to it. Then, add a Create Random Points node (Node 10) and link Parallel Chunk Start (Node 9) to it. In the dialog of Create Random Points (Figure 12.2), choose Popu for Number of points column and BLOCKID10 for ID column, and click OK. Add a Parallel Chunk End node (Node 11), and link Node 10 to it. Keep the default settings in the dialog of Parallel Chunk Start (Node 9) (Figure 12.3a), and check Add Chunk Index to RowID in the dialog of Parallel Chunk End (Node 11) (Figure 12.3b) to avoid potential conflicts of row IDs. Finally, execute the three nodes to generate random points.[1]

The resulting data have the same number of rows as the input data. Use a Multipart to SinglePart node (Node 12) to convert the points in a MultiPoints format to single points.

Step 3. Designing a uniform analysis area unit: Add a GeoFile Reader node (Node 13) to the Editor and input the file path of census tracts (e.g., E:\KNIME\CMGISV3\ Chp12\data\tract.zip) to read the data. Add a Column Filter node (Node 14) to keep the key columns OBJECTID and geometry.

Under Spatial Manipulation, drag a Create Grid node (Node 15) to the Editor, and link Column Filter (Node 14) to it. In the dialog, input 1000 for Grid Length. The

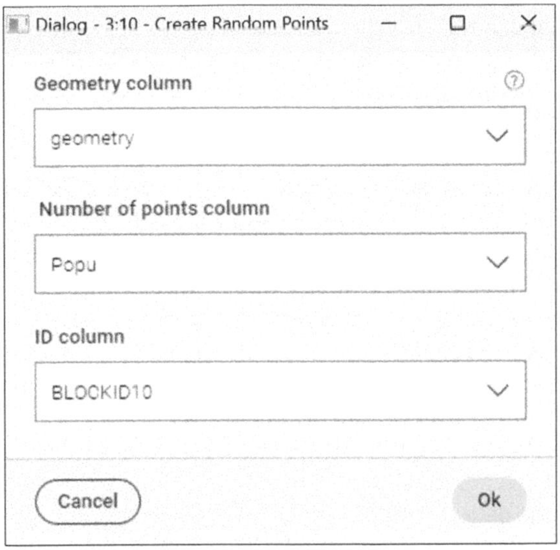

FIGURE 12.2 Interface of Create Random Points

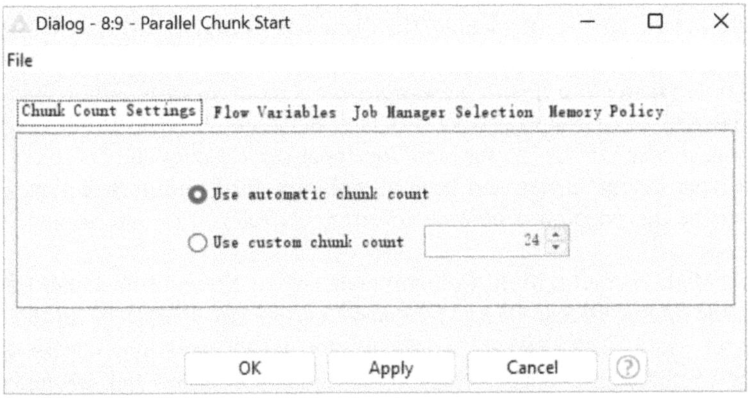

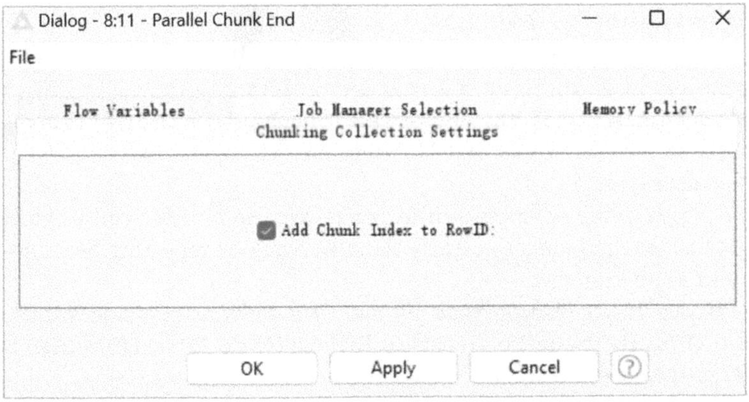

FIGURE 12.3 Interfaces of (a) Parallel Chunk Start, and (b) Parallel Chunk End

result is a grid of 1km × 1km with an ID column `Grid ID` according to the extent and CRS of the input data.

Step 4. Aggregating points (population) by grids: Add a Spatial Join node (Node 16) and connect Multipart to SinglePart (Node 12) and Create Grid (Node 15) to it. Execute it with the default settings, while setting `Intersects` as Match mode and `Inner` as Join mode.

Add a GroupBy node (Node 17) to summarize the `Count` of the column `BLOCKID10` by the group column `Grid ID`, while keeping `Aggregation method(column name)` for Column naming. The resulting column `Count*(BLOCKID10)` is the total point counts (or population) for each grid (1 square kilometer).

Step 5. Measuring distances between grid centroids and the city center: Add a Geometry To Point node (Node 18) and link Create Grid (Node 15) to it. In the dialog, choose `centroid` and check `Replace` for Output column in the dialog. Execute it to transform the grids into their centroids.

Add a GeoFile Reader node (Node 19) to the Editor and input the file path of city center (e.g., `E:\KNIME\CMGISV3\Chp12\data\citycenter.zip`). Execute the node to read the data.

Add a Euclidean Distance node (Node 20), and link Geometry To Point (Node 18) and GeoFile Reader (Node 19) to it. In the dialog, choose `Grid ID` for Origin ID column, `ORIG_FID` for Destination ID column, `Kilometer` for Distance unit, click OK and execute it to calculate the distances between grid centroids and the city center.

Step 6. Implementing linear regressions (grid as area unit): Add a Joiner node (Node 21), and link GroupBy (Node 17) and Euclidean Distance (Node 20) to it. In the dialog, choose `Grid ID` for both Top Input ('left' table) and `Origin ID` for Bottom Input ('right' table), and keep all columns for Column Selection. Execute it to combine the population column `Count*(BLOCKID10)` and distance column `Distance`.

Add a Math Formula (Multi Column) node (Node 22) and link Joiner (Node 21) to it. In the dialog, choose `Distance` and `Count*(BLOCKID10)` in the Include box, input `ln($$CURRENT_COLUMN$$)` for the Expression, check `Append Selected Columns with Suffix` and input `_ln` for it. Click OK and execute it to apply a logarithmic operation on the two columns and save them as two new columns, `Distance_ln` and `Count*(BLOCKID10)_ln`.

Use a Linear Regression Learner node (Node 23) to explore any of the four regressions of `Distance` (or `Distance_ln`) vs. `Count*(BLOCKID10)` (or `Count*(BLOCKID10)_ln`) as discussed in Chapter 6. Note that the areal unit is grid with a uniform area size of 1 km² and thus its population is equivalent to its population density.

Step 7. Aggregating points by census tracts: We can also use census block group or census tract as area units to explore the difference by repeating Steps 4–6. Take census tract as an example.

Copy Nodes 16–21 (except Node 19) and their links as new nodes: Spatial Join (Node 24), GroupBy (Node 25), Geometry To Point (Node 26) and Euclidean Distance (Node 27) and Joiner (Node 28).

Link Multipart to SinglePart (Node 12) and Column Filter (Node 14) to it. In the dialog, choose `Inner` for Join mode, `Intersects` for Match mode, click OK and execute it to attach the `OBJECTID` of each census tract to each generated random point.

Use GroupBy (Node 25) to summarize the total number of points. In the dialog, (1) under the Groups tab, choose `OBJECTID` for Group column (s) and `Aggregation method(column name)` for Column naming, (2) under Manual Aggregation tab, choose `BLOCKID10` for Column and `Count` for Aggregation. Click OK and execute it. In the result table, the field `Count*(BLOCKID10)` represents the total number of points (population) in each census tract.

Step 8. Measuring distances between census tract centroids and the city center: Link Column Filter (Node 14) to Geometry to Point node (Node 26). Execute it to transform the data into centroid points.

Link Geometry to Point (Node 26) and GeoFile Reader (Node 19) to Euclidean Distance node (Node 27). Execute it to compute Euclidean distances between tract centroids and the city center. In the output table, column `Origin ID` is for census tract centroids, and column `Destination ID` is for the city center.

Link GroupBy (Node 25) and Euclidean Distance (Node 27) to Joiner (Node 28). In the dialog of Joiner (Node 28), choose `OBJECTID` for Top Input ('left' table) and `Origin ID` for Bottom Input ('right' table), and keep all columns. Execute it to attach the point counts (simulated population) to the OD list.

Add a Joiner node (Node 29), and link GeoFile Reader (Node 13) and Joiner (Node 28) to Joiner (Node 29). In the dialog of Joiner (Node 29), choose `OBJECTID` for both Top Input ('left' table) and Bottom Input ('right' table); and under Column Selection, choose `Shape_Area` for the Include box for Top Input (left table) and keep all columns in Bottom Input (right table). Execute it to attach the area.

Finally, add a Math Formula node (Node 30) to append a column `PopuDen` with the expression `$Count*(BLOCKID10)$*1000000/$Shape _ Area$`. The resulting column `PopuDen` represents population densities in census tracts. Unlike the area unit grid with a uniform area of 1 square kilometer, here we incorporate area size to calculate population densities in census tracts.

Step 9. Implementing linear regressions (census tract as area unit): Add a Math Formula (Multi Column) node (Node 31) and link Math Formula (Node 30) to it. In the dialog, choose `Distance` and `PopuDen` in the Include box, input `ln($$CUR-RENT_COLUMN$$)` for the Expression, check `Append Selected Columns with Suffix` and input `_ln` for it. Click OK and execute it to apply a logarithmic operation on the two columns and save them as two new columns, `Distance_ln` and `popuDen_ln`.

Use a Linear Regression Learner node (Node 32) to explore any of the four regressions of `Distance` (or `Distance_ln`) vs. `PopuDen` (or `PopuDen_ln`).

Step 10. Comparing units in sample distribution across distance ranges: Add a Math Formula node (Node 33) and a GroupBy node (Node 34), and link Joiner (Node 21) → Math Formula (Node 33) → GroupBy (Node 34).

Configure Math Formula (Node 33) to append a new Integer column `DistClass` with the expression `ceil($Distance$/10)`. This classifies the distances with equal intervals of 10 km. In the dialog of GroupBy (Node 34), (1) under the Groups tab, choose `DistClass` for Group column(s) and `Aggregation method(column name)` for Column naming, (2) under Manual Aggregation tab, choose `Grid ID` for Column and `Count` for Aggregation. Click OK and execute it. In the result table, the field `Count*(Grid ID)` represents the number of census tracts by intervals.

Copy the two nodes Math Formula (Node 33) and GroupBy (Node 34) as two new nodes, and link Joiner (Node 29) → Math Formula (Node 35) → GroupBy (Node 36). Keep the predefined setting in Math Formula (Node 35) to append a new column `DistClass`. In the dialog of GroupBy (Node 36), set `OBJECTID` for Column and `Count` for Aggregation under Manual Aggregation tab. Similarly, the field `Count*(OBJECTID)` represents the number of census tracts by intervals.

Add a Joiner node (Node 37) and link two GroupBy nodes (Nodes 34 and 36) to it. In the dialog, choose `DistClass` for both Top Input ('left' table) and Bottom Input ('right' table). Execute it to join the two tables.

Use a Number to String node (Node 38) to convert `DistClass` as a string column. Add a Bar Chart node (Node 39), and link Number to String (Node 38) to it. In the dialog shown in Figure 12.4, set `DistClass` as Category Column, check Average for Aggregation Method, then choose `Count*(Grid ID)` and `Count*(OBJECTID)` in the box. Execute it to show the changes of grids and tracts along the distances to CBD.

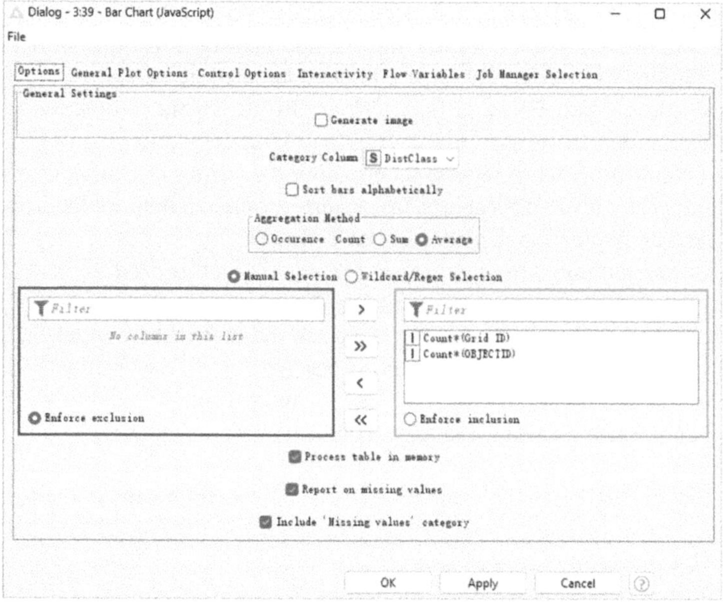

FIGURE 12.4 Interface of Bar Chart

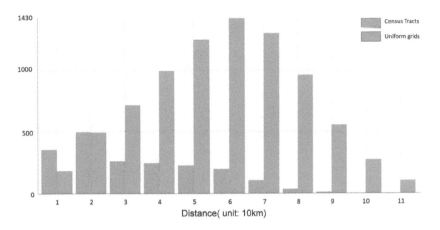

FIGURE 12.5   Numbers of census tracts and uniform grids by distance ranges

As shown in Figure 12.5, the number of census tracts generally declines with increasing distance from the city center, and it suggests oversampling in near distance ranges; however, the distribution of grid count largely conforms to a normal distribution, which is more desirable when applying OLS regression analysis.

12.2   CASE STUDY 12B: MONTE CARLO BASED TRAFFIC SIMULATION IN BATON ROUGE, LOUISIANA

This case study is developed to illustrate the application of Monte Carlo simulation in traffic simulation. The study focuses on East Baton Rouge Parish (EBRP) of Louisiana, the urban core of Baton Rouge MSA, but extends to include its eight neighboring parishes in traffic simulation in order to account for internal-external, external-internal, and through traffic.

The subfolder BRMSAmc under the data folder BRMSA includes:

1) a zipped area feature trt.zip and its corresponding point feature trtpt.zip for 151 census tracts and their centroids, respectively (each containing fields res and emp for numbers of resident workers and jobs, respectively),

2) a zipped road feature dataset rd.zip containing all feature classes associated with the road network dataset,

3) a zipped point feature station.zip for 816 traffic count stations (its field ADT1 representing the annual average daily traffic or AADT), and

4) Python files, ProbSampling.py, RepeatRow.py and Shortest Network Paths.py for sampling with probability, repeating rows, and calculating shortest paths, respectively.

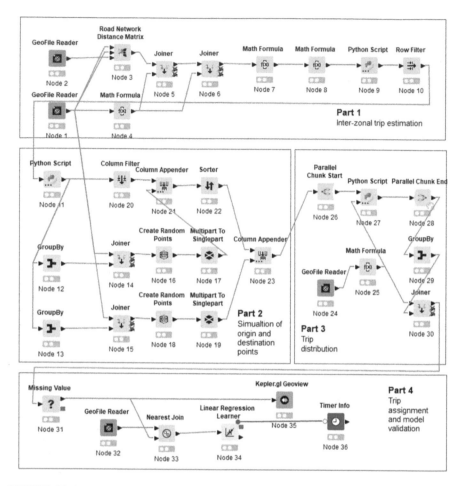

FIGURE 12.6 Workflow of Case Study 12B

Figure 12.6 shows the workflow of Case Study 12B. Part 1 uses a gravity model to estimate the zone-level traffic between tracts (steps 1–2). Part 2 uses the Monte Carlo method to randomly simulate individual trip origins and destinations, proportional to the distributions of resident workers and employment, respectively (steps 3–4). Part 3 connects the origins and destinations randomly and caps the volume of O-D trips proportionally to the estimated inter-zonal traffic from Part 1 (step 5). Part 4 calibrates the shortest routes for all O-D trips, measures the simulated traffic through each monitoring station and compares it to the observed traffic for validation (steps 6–8).

As shown in Figure 12.7, a sample dataset of four units is designed to illustrate the procedures in Parts 1–3.

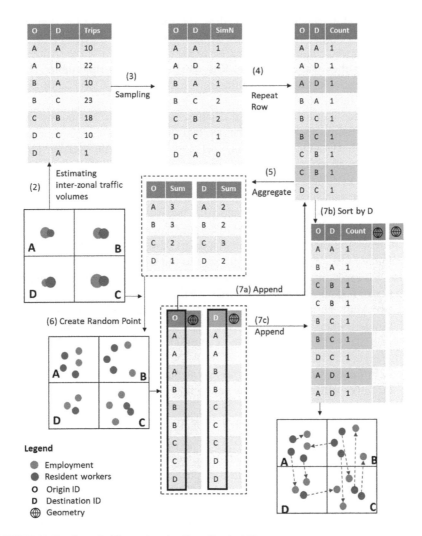

FIGURE 12.7 Sample illustration for Case Study 12B

12.2.1 PART 1: ESTIMATING INTER-ZONAL O-D TRIP VOLUMES

Step 1. Computing OD time matrix between census tracts: In KNIME AP, create a new workflow `Case12B`. Add two GeoFile Reader nodes to the Editor and input the file paths of census tract units in Node 1 (e.g., `E:\KNIME\CMGISV3\Chp12\data\trt.shp`) and road network in Node 2 (e.g., `E:\KNIME\CMGISV3\Chp04\data\rd.zip`), and execute them to read the data.

Add a Road Network Distance Matrix node (Node 3). Link GeoFile Reader (Node 1) to its first two input ports as both origin and destination, and link GeoFile Reader

(Node 2) to its third input port for feeding the road network. In the dialog, input `trtID` for both Origin ID column and Destination ID column, choose SPEED for Road network speed column from road network, check `Miles per hour`(mph) for Speed unit, click OK and execute it to calculate the OD time table. The table displays the duration of each trip in minutes. There are $151 \times 151 = 22,801$ OD pairs in the resulting table.

Similar to step 2 in Section 10.3 of Chapter 10, we need to amend the OD time table by adding intra-zonal times at the two trip ends.

Add a Math Formula node (Node 4) and link GeoFile Reader (Node 1) to it. Configure it to append a new column `intrazone` with the expression `0.25*$Length$/670.56`, which is approximated as ¼ of a tract's perimeter divided by a constant speed (670.56 m/min, i.e., 25 mph).

Add two Joiner nodes (Nodes 5 and 6). Link Road Network Distance Matrix (Node 3) and Math Formula (Node 4) to Joiner (Node 5), and link Joiner (Node 5) and Math Formula (Node 4) to Joiner (Node 6). Use the two nodes to append the three columns, `emp`, `res` and `intrazone` to the OD list by the Join columns of `Origin ID` (or `Destination ID`) and `trtID`, respectively. For a duplicated column name, it is appended with a suffix (#1).

Add two Math Formula nodes (Nodes 7 and 8) and link Joiner (Node 6) → Math Formula (Node 7) → Math Formula (Node 8). Configure Node 7 to update the column `Duration` using the expression `$Duration$+$intrazone (right)$+$intrazone$`.

Step 2. *Estimating inter-zonal traffic volumes*: Recall the gravity model for estimating the daily traffic between zones *i* and *j* is:

$$T_{ij} = a\left(W_i E_j + E_i W_j\right)d_{ij}^{-\beta} \tag{12.1}$$

where *W* and *E* are resident works and employment, respectively and *D* is distance or travel time between them. We begin with parameter settings in equation (12.1) as $\beta = 2$ and $a = 0.001$.

To estimate the inter-zonal traffic volumes, configure Math Formula (Node 8) to append a new integer column `Trip` with the expression `0.001*($emp$*$res (right)$+$res$ *$emp (right)$) *$Duration$^(-2)`. The sum on the field Trip is 323,629. This step is illustrated as step (2) in Figure 12.7.

Step 3. *Monte Carlo simulation of OD traffic*: Due to computation limitations, our simulation is downscaled to 75,000 trips. In other words, the simulated OD traffic volumes will be proportionally to the numbers of estimated trips (column `Trip` in Node 8) at a ratio of 75000/323629.

A Python script file `ProbSampling.py` based on the *np.random* function in the Numpy package is used to implement a Monte Carlo simulation of OD traffic volumes according to the distribution of `Trip`. Add a Python Script node (Node 9) and link Math Formula (Node 8) to it. In the coding box, input the codes as in Table 12.1, and execute it.

The resulting table has a column `SimN` indicating the number of simulated trips. The values of `SimN` and `Trip` are highly correlated. Use a Row Filter node

TABLE 12.1
Python Code for Sampling with Probability

	Scripts	Explanation
1	import knime.scripting.io as knio import numpy as np import collections import pandas as pd	Import packages
2	df=knio.input_tables[0].to_pandas().reset_index(drop=True)	Import data & reset index
3	df['prob']=df.Trip/sum(df.Trip)	Calculate Probability
4	num=75000	Total trips
5	simlist=np.random.choice(range(0,df.shape[0]), num, p=df.prob).tolist()	Simulation by probability
6	ct = collections.Counter(simlist)	Count frequency
7	ct_df = pd.DataFrame.from_dict(ct, orient='index').reset_index()	Save as data frame
8	ct_df = ct_df.rename(columns={0:'SimN'})	Rename as SimN
9	mg=pd.merge(df.reset_index(),ct_df, on='index',how='left')	Join table
10	mg['SimN'].fillna(0, inplace = True)	Fill NA value
11	knio.output_tables[0] = knio.Table.from_pandas(mg)	Export data

(Node 10) to keep only the records with SimN ≥ 1. This step is illustrated as step (3) in Figure 12.7.

This completes Part 1 for estimating inter-zonal O-D trip volumes.

12.2.2   PART 2: SIMULATING INDIVIDUAL TRIP ORIGINS AND DESTINATIONS

Step 4. Disaggregating inter-zonal OD traffic flows to individual OD pairs: The OD list table from step 3 contains three key fields: Origin ID, Destination ID and SimN. To generate individual trips, each row in the table must be repeated as many times as the inter-zonal trip count. A Python script file RepeatRow.py (Node 11) based on *pd.DataFrame.index.reperat* is used to repeat rows by the value of SimN. Add a new Python Script node (Node 11), link Row Filter (Node 10) to it, and input the codes in Table 12.2. The resulting table has 75,000 rows for the trips to be simulated, with a column Count = 1 for each row. This step is illustrated as step (4) in Figure 12.7.

Step 5. Calibrating numbers of simulated origin and destination points in census tracts: Add two GroupBy nodes (Nodes 12 and 13) and link Python Script (Node 11) to them. Use Nodes 12 and 13 to aggregate the Sum for the column Count by the Group column Origin ID and Destination ID, respectively. Remember to choose Keep original name(s) for Column naming in the dialogs. The resulting column Count represents the number of resident workers

TABLE 12.2
Python Code for Repeating Rows

	Scripts	Explanation
1	import knime.scripting.io as knio import numpy as np import pandas as pd	Import packages
2	df=knio.input_tables[0].to_pandas().reset_index(drop=True)	Import data & reset index
3	df["SimN"] = df["SimN"].astype(int)	Assure Integer column
4	dfnew=df.loc[df.index.repeat(df.SimN)]	Repeat row
5	dfnew["Count"]=1	Add count
6	dfnew.reset_index(drop=True,inplace=True)	Reset index
7	knio.output_tables[0] = knio.Table.from_pandas(dfnew)	Export data

(and jobs) in each census tract to be simulated. This step is illustrated as step (5) in Figure 12.7.

Step 6. Monte Carlo simulation of individual origins and destinations: Add two Joiner nodes (Nodes 14 and 15), link GeoFile Reader (Node 1) to the first input port of Joiner (Nodes 14 and 15), link GroupBy (Node 12) to the bottom input port of Joiner (Node 14), and link GroupBy (Node 13) to the bottom input port of Joiner (Node 15). Implement the attribute join based on their common fields trtID, Origin ID and Destination ID, and only keep the column geometry in the Top input (left table) box of both nodes. The two resulting tables contain the key columns, Count, geometry and Origin ID (or Destination ID). The column Count indicates the number of points to be simulated.

Add two Create Random Points nodes (Nodes 16 and 18) and two Multipart to Singlepart nodes (Nodes 17 and 19). Link Joiner (Node 14)→ Node 16→ Node 17, and link Joiner (Node 15) → Node 18→ Node 19. In the dialogs of Create Random Points nodes, input Count for Number of points column and Origin ID (or Destination ID) for ID column. Click OK and execute the nodes to generate points in the census tracts as origins and destinations. Execute the Multipart to Singlepart nodes to convert the MultiPoint geometries into individual points. This step is illustrated as step (6) in Figure 12.7.

This completes Part 2 for randomly simulating individual trip origins and destinations.

12.2.3 PART 3: SIMULATING INDIVIDUAL OD PAIRS

Step 7. Appending simulated origins and destinations to individual OD pairs: The number of origin (or destination) points sharing a certain ID column is identical to the number of trips with a corresponding Origin ID (or Destination ID) in the table of Python Script (Node 11). Such a relationship is utilized to randomly assign the points to trips.

Add a Column Filter node (Node 20) and link Python Script (Node 11) to it. Configure it to keep the key columns, `Origin ID` and `Destination ID`. Add a Column Appender node (Node 21), and link Column Filter node (Node 20) and Multipart to Singlepart (Node 17) to it. Execute it to combine the two tables (remember to check `Generate new row keys`). This assigns the simulated points as origin points in the OD list. This step is illustrated as step (7a) in Figure 12.7.

As the resulting table is by default sorted in accordance with the rank of `Origin ID`, we need to resort it with `Destination ID` to append the destination points. Add a Sorter node (Node 22) and link Column Appender (Node 21) to it. In the dialog (as shown in Figure 10.8), choose `Destination ID` for Sort by and check `Ascending`. Execute it to resort the table. This step is illustrated as step (7b) in Figure 12.7.

Add another Column Appender node (Node 23), and link Sorter (Node 22) and Multipart to Singlepart (Node 19) to it. Execute it to append the destination points to the OD list. The resulting table has 75,000 rows, and each row identifies a unique OD pair. This step is illustrated as step (7c) in Figure 12.7 and 12.8.

FIGURE 12.8 Interface of Sorter

This completes Part 3 for connecting the origins and destinations randomly to form unique OD pairs.

12.2.4 PART 4: TRIP ASSIGNMENT AND MODEL VALIDATION

Step 8. Computing shortest paths between origins and destinations: This step uses the Parallel Execution extension again for computational efficiency and a Python Script node to compute the shortest paths.

Copy the GeoFile Reader node (Node 2) as a new node GeoFile Reader (Node 24) and execute it to load the data again. Use a Math Formula node (Node 25) to append a new integer column RID with the expression $$ROWINDEX$$+1.

Add a Parallel Chunk Start node (Node 26) and link Column Appender (Node 23) to it. Add a Python Script node (Node 27) and configure it to have two input ports, and link Parallel Chunk Start node (Node 26) and Math Formula (Node 25) to it. In the dialog, input the script (provided in Shortest Network Paths.py) from Table 12.3 in the coding box. After that, add a Parallel Chunk End node (Node 28) and

---

TABLE 12.3
Python Code for Calculating Shortest Paths

	Scripts	Explanation
1	import knime.scripting.io as knio import geopandas as gp import pandas as pd import numpy as np import momepy # urban network analysis from scipy.spatial import cKDTree import networkx as nx # urban network analysis	Import packages
2	origin_gdf=gp.GeoDataFrame(knio.input_tables[0].to_pandas() [['geometry']], geometry='geometry')	Import data as origin GeoDataFrame
	destination_gdf=gp.GeoDataFrame(knio.input_tables[0].to_pandas() [['geometry (#1)']], geometry='geometry (#1)')	Import destination data
	destination_gdf.rename(columns={"geometry (#1)":"geometry"}, inplace=True) destination_gdf=destination_gdf.set_geometry("geometry")	Revise geometry name
	road=gp.GeoDataFrame(knio.input_tables[1].to_ pandas(),geometry='geometry')	Import road data
3	graph = momepy.gdf_to_nx(road, approach='primal')	Create graph
	nodes, edges = momepy.nx_to_gdf(graph, points=True, lines=True, spatial_weights=False)	Rebuild nodes and edges

Scripts	Explanation

4 `graph_ID = nx.Graph()`
`for i,row in edges.iterrows():`
`  node_start = row['node_start']`
`  node_end = row['node_end']`
`  routeID = row['RID']`
`  weight=row['Minutes']`
`  graph_ID.add_edge(node_start,node_end,routeID=routeID)`
	Generate graph indexed by ID

5 `def ckdnearest(gdA, gdB):`
`  nA = np.array(list(gdA.geometry.apply(`
`  lambda x: (x.x, x.y))))`
`  nB = np.array(list(gdB.geometry.apply(`
`  lambda x: (x.x, x.y))))`
`  btree = cKDTree(nB)`
`  dist, idx = btree.query(nA, k=1)`
`  gdB_nearest = gdB.iloc[idx].reset_index(drop=True)`
`  gdf = pd.concat([gdA.drop(`
`  columns="geometry").reset_index(drop=True),`
`  gdB_nearest,`
`  pd.Series(dist, name='dist')],`
`  axis=1)`
`  return gdf`
	Define function of nearest points
	Get the coordinates
	Search nearest nodes in Graph
	Calculate distance

6 `originNearestPoint = ckdnearest(origin_gdf, nodes)` `destinationNearestPoint = ckdnearest(destination_gdf, nodes)`	Find nearest nodes in graph
7 `x = originNearestPoint['nodeID'].to_numpy()` `y = destinationNearestPoint['nodeID'].to_numpy()`	Expand NodeID as list

8 `pathList = []`
`for i in range(x.size):`
`  fromNode = x[i]`
`  toNode = y[i]`
`  try:`
`  sp = nx.shortest_path(graph_ID, fromNode, toNode,`
`  method='dijkstra',weight="weight")`
`  pathGraph = nx.path_graph(sp)`
`  for ea in pathGraph.edges():`
`  pathList.append({'origin': fromNode,`
`  'destination': toNode,'fromJuction': ea[0],`
`  'toJuction': ea[1],`
`  'RID': graph_ID.edges[ea[0], ea[1]]['routeID']})`
`  except nx.NetworkXNoPath:`
`  pass`
	Calculate shortest path

9 `route_df = pd.DataFrame(pathList)`	Save as table
10 `knio.output_tables[0] = knio.Table.from_pandas(route_df)`	Export data

link Python Script (Node 27) to it. In the dialog of Parallel Chunk Start (Node 26), input 6 (or a smaller number[2]) for Use custom chunk count to mitigate surging computation resource need by loading the road network. In the dialog of Parallel Chunk End (Node 28), check Add Chunk Index to RowID to avoid potential conflicts of row IDs.

Execute the three nodes to calculate the shortest paths between the simulated origins and destinations. The result table contains five columns for all road segments in the shortest paths: origin, destination, fromJunction, toJunction and RID.

Step 9. Aggregating road segments on the shorted paths: Add a GroupBy node (Node 29) and link Parallel Chunk End (Node 28) to it. Configure it to aggregate the Count for the column origin based on Group column RID, while checking Aggregation method(column name) for Column naming. Execute it. Column Count*(origin) records the traffic volume on each road segment.

Add a Joiner node (Node 30), and link Math Formula (Node 25) and GroupBy (Node 29) to it. In the dialog, (1) under Joiner Settings tab, choose RID for both Top Input ('left' table) and Bottom Input ('right' table); (2) under Include in output, check Matching row and Left unmatched row to apply a left join; (3) under Column Selection, keep all columns in the Include box for Top Input (left table) and choose Count*(origin) for Bottom Input (right table); (4) keep other default settings, and execute it. This step attaches the traffic volumes to the road network.

The result table may contain empty values that indicate no traffic records on those roads. Add a Missing Value node (Node 31) and link Joiner (Node 30) to it. In the dialog, set Fix Value for Number (integer) and input 0 for it. Click OK and execute it to replace all null values as 0.

Step 10. Comparing simulated and observed traffic patterns: Add a GeoFile Reader node (Node 32) to the Editor and input the file path of traffic monitoring stations (e.g., E:\KNIME\CMGISV3\Chp12\data\station.zip). Execute it to read the data.

Add a Nearest Join node (Node 33). Link GeoFile Reader (Node 32) and Missing Value (Node 31) to it. In the dialog, choose Inner for Join mode, 1000 for Maximum distance, and Meter for Distance unit. Execute it to attach the attributes of the nearest road to the stations.

Then, we can use a Linear Regression Learner node (Node 34) to explore the correlation between ADT1 (average daily traffic) and Count*(origin) (simulated traffic) at traffic monitoring stations. Figure 12.9 uses a Kepler.gl Geoview node (Node 35) to visualize the traffic volume along the roads. Additionally, a Timer Info node (Node 36) can be added to collect execution time information and facilitate workflow improvement.

To examine the impact of distance decay coefficient parameter β in Equation (12.1) on the model, we can use a Counting Loop Start node (Node 36) and Math Formula (Variable) node (Node 37) to build a varying parameter β (e.g., 1.2–2.8) and a loop workflow to compare the results. Figure 12.10 shows that $\beta = 2$ produces the best fitting power. See the loop workflow in the data folder.

FIGURE 12.9   Simulated traffic volumes

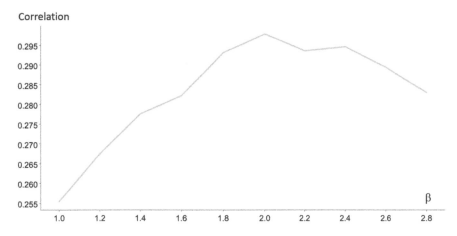

FIGURE 12.10   Correlations between observed vs simulated traffic with various β values

12.3   CONCLUDING REMARKS

Chapter 12 discusses the Monte Carlo simulation technique. The Geospatial Analytics Extension's Create Random Points node plays a crucial role in simulations. Three new Python Script nodes are introduced: sampling with probability, repeating rows, and calculating shortest paths. The KNIME extension of Parallel Execution applies multiple threading computation in the workflow to accelerate data processing. This feature is particularly useful for intensive computation such as simulation tasks.

A helpful tip is that converting numeric values to string values enables the bar chart to display multiple values simultaneously.

NOTES

1 It takes several minutes on our laptop.
2 The Python Script node processes the road data consisting of 76,574 rows. Given the large number of parallel processes, the memory usage may become unpredictable.

13 Agent-Based Model and Application in Crime Simulation

This chapter furthers the endeavor of simulation in Chapter 12 to model individual spatiotemporal trajectories by *agent-based modeling (ABM)*. Simulated agents (various groups of individuals) guided by defined rules (routines) move in space (a study area) over time, and their convergence in both space and time predicts the likelihood for an event to take place. In a way, an agent-based model mimics human movements in real life. When the functions and parameters used in the model are well informed from increasingly detailed real-world empirical data (e.g., big data discussed in Chapter 14), the model has great potentials for capturing the dynamics of human-environment and human-to-human interactions and thus explaining and predicting spatial patterns and underneath processes more accurately.

Case Study 13 uses ABM to simulate the spatiotemporal trajectory of robbery in Baton Rouge. It highlights the role of ABM as an artificial laboratory for social science experiments by simulating various scenarios to explore crime theory and alternative policing strategies. Simulation runs are compared by their abilities to predict crime hotspots from actual data. For each simulation run, a series of kernel density estimate (KDE) maps are created from simulated robbery events. Areas with crime density above two standard deviations are treated as predicted crime hotspots from the corresponding simulation run. The predictive ability of crime hotspots from simulation is evaluated with three metrics: *hit rate (H)*, *prediction accuracy index (PAI)*, and *forecast accuracy index (FAI)*, as discussed in the main book.

This case study uses GIS and tabular data in traditional ESRI shapefile (.shp), dBASE (.dbf), and comma-separated values (.csv) formats. These formats were chosen for their better compatibility with the open-source Python GIS libraries used to build the core of this simulation model, including *GDAL*, *Shapely* and *NetworkX*.

Major files in the subfolder `Data` under the folder `ABMSIM` are organized as below:

1. Under the subfolder `ebr_grid`, (1) `ebr_grid_500m_with_roads.shp` is a spatial grid layer covering the entire study area with a 500-meter resolution, and (2) `np_odmatrix.npy` is an origin-destination travel distance matrix between grid centers. Since resident agents can only travel on roads, only those grids with roads inside are kept for the spatial index layer. They are used by agents in heuristic estimates on travel costs.
2. In the root directory, (1) `ebr_boundary.shp` represents the boundary of East Baton Rouge Parish, (2) `ctpp_ebr_taz_utm_fixed.shp` provides the commuting flows between TAZs extracted from the CTPP

(Census Transportation Planning Products) data, (3) `ebr_taz_flow_ total_workers_sparse.dbf` records the worker flows for TAZs, (4) `tl_2015_22033 _road_segments _utm.shp` is the line layer for the road network, (5) `ebr_random_points_ grid_500m_taz.shp` is a point layer with all randomly generated points on road network, and (6) `ebr_acs_2015_5yr_bg.shp` is the census block group layer.

3. In the root directory, the file `ABMCrimePara.csv` contains all the parameters used in the model. It has five columns: `Variable` (flow variable names used in KNIME), `Value` (default values of variables), `Explanation` (detailed information), `Category` (variable types) and `ArcToolOrder` (corresponding order in the toolbox ABM Crime Simulator for `ArcGIS Pro.tbx` for reference). `ebr_2011_robbery. shp` represents 2011 reported robberies with a column `OffenseMon` for months.

4. Python file `ABMRouteKNIME.py` is the main script used in the workflow, where its dependent files are in the sub folder `ABMscripts`, and `ABM-Kernel.py` is used for implementing Kernel density estimation.

5. The result data `sim_crime_incidents.csv` from step 4 in Part 1 is provided for users to conveniently explore the experiments in Part 2.

Refer to the Main book for more detailed information on the model input data in simulation mode.

13.1 PART 1: SIMULATING CRIMES IN THE BASE SCENARIO IN ABM IN BATON ROUGE

The case study begins with simulating crimes in the base scenario. This sets up the baseline for testing some hypotheses suggested in some crime theory and policing strategy. Figure 13.1 shows the workflow for Part 1 of Case Study 13.

Step 1. Preparing running parameters for the ABM simulation: To begin, launch the KNIME Analytics Platform and create a new workflow `Case13`. Add a CSV Reader node (Node 1) to the Editor and input the file path for simulation parameters, `ABMCrimePara.csv`, e.g., `E:\KNIME\CMGISV3\Chp13\ABMCrimePara. csv`. In the dialog, select `UTF-8` in the Encoding tab. Click OK and execute the node to read the data. The ABM model uses 59 parameters, as indicated by the 59 rows in the result table.

Step 2. Adjusting the file path for Python workspace: To ensure that the Python script can access the necessary input and output data, we need to modify the file paths in the simulation parameters. These parameters use relative paths, e.g., `ABMSIM\ Data\ebr_boundary.shp`, and we need to append the local path to create a full path, such as `E:\CMS_GISV3\Chp13\ABMSIM\Data\ebr_boundary.shp`. This is implemented as follows.

Add a Row Splitter node (Node 2) and link it to the CSV Reader node (Node 1) that reads the simulation parameters. In the dialog, (1) check `Include rows by`

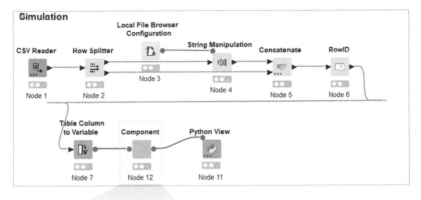

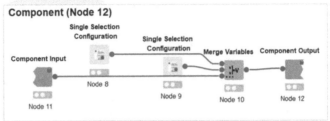

FIGURE 13.1 Workflow for Part 1 of Case Study 13

attribute value, choose Value for Column to test, (2) under Matching criterion, check use pattern matching and input ABMSIM.* for it, and check regular expression. Click OK and execute it. It exports the variables related to file paths to the upper output port.

Add a Local File Browser Configuration node (Node 3). In the dialog shown in Figure 13.2, input the folder where the ABMSIM folder is located, such as, e.g., E:\CMS_GISV3\Chp13, keep other default settings and ignore the warning of "Errors:input location is a directory". Click OK and execute it. It creates a flow variable file-input with the string value of "E:\CMS_GISV3\Chp13".

Add a String Manipulation node (Node 4). Link Local File Browser Configuration (Node 3) to the flow variable port of Node 4, and link the upper output port of Row Splitter (Node 2) to the input data port of Node 4, respectively. Configure it to replace the value in the column Value with the expression join($$\{Sfile-input}$$,"\\",$Value$). This step appends the local path to all relative paths in the table.

Add a Concatenate node (Node 5) and link the String Manipulation node (Node 4) and the bottom output port of the Row Splitter node (Node 2) to it. Execute this node to combine the table vertically. Finally, use a RowID node (Node 6) to replace the row IDs with the column Variable, and remember to check "remove the selected column".

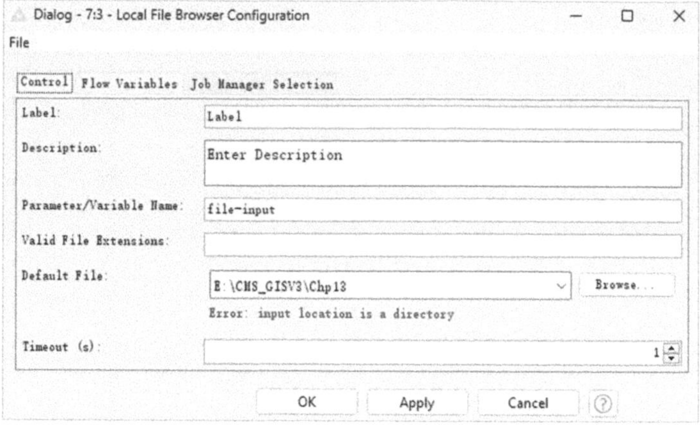

FIGURE 13.2   Interface of Local File Browser Configuration

Step 3. Adjusting parameters for simulation: Add a Table Column to Variable node (Node 7) and link RowID (Node 6) to it. Execute it directly with the default setting, which uses the first column `Value` for the variables derived from row ID names. It generates a series of flow variables, whose names are the row ID values derived from the previous column `Variable` with the values coming from the first column `Value`.

There are multiple ways to modify parameter values in the simulation model. One way is to edit the source table of `ABMCrimePara.csv` directly. Another option is to create a copy of the source table using a Table Creator node and modify the values as needed. Alternatively, flow variable nodes can be used to overwrite default values. Here, we use Single Selection Configuration nodes to modify the default values for the simulation model.

Add two Single Selection Configuration nodes (Nodes 8 and 9). In the dialog of Node 8, input `Apply journey-to-crime behavior` for Label and `offender_j2c_behavior` for Parameter/Variable Name, input `true` and `false` in two lines for Possible Choice, choose `false` in the Default Value, and click OK and execute it. This creates a flow variable `offender_j2c_behavior` with the default value false. One may choose `true` to apply journey-to-crime behavior.

In the dialog of Node 9, input `Crime opportunity field in spatial grid layer` for Label and `offender_opportunity_field` for Parameter/Variable Name, input `CRIME2010` and `UTILITY` in two lines for Possible Choice, choose `UTILITY` in the Default Value, click OK and execute it. It creates a flow variable `offender_opportunity_field` with the default value `UTILITY`. Both `CRIME2010` and `UTILITY` are fields of `ebr_grid_500m_with_roads.shp`.

Add a Merge Variables node (Node 10), configure it to have three input ports, and link Table Column to Variable (Node 7) and two Single Selection Configuration nodes (Nodes 8 and 9) to it. Add a Python View node (Node 11) and link the output port of Merge Variables (Node 10) to it.

FIGURE 13.3   Interface of Component for parameter adjustment

Select Nodes 8, 9, and 10, and convert them into a Component (Node 12) by right clicking any of the nodes and choosing Create Component. Right-click the component > Open. In the Editor of Component (Node 12) as shown in the bottom part of Figure 13.1, run the whole sub-workflow. Double-click the Component Input (Node 11) to bring out the dialog, exclude the two flow variables `offender_j2c_behavior` and `offender_opportunity_field` from the Include box wherein all variables are from Table Column to Variable (Node 7). In the dialog of Component Output (Node 12), keep all variables in the include box.

We can use more flow variable nodes to enrich the parameter options in the component. Figure 13.3 shows the Interface of Component for parameter adjustment.

Step 4. Running simulation with flow variable and Python View node: The Python Script node (Node 11) is prepared to run the simulation. In the dialog box, copy the script from `ABMRouteKNIME.py` and paste it into the coding box. Then click OK to run the script.

The result of simulated crime incidents with the default parameter settings is stored in the file `sim_crime_incidents.csv`. This CSV file, along with the other two resulting files such as `sim_agent_location_history.csv` and `sim_agent_type_home_work.csv` can be used for animation and further analysis.

13.2   PART 2: EVALUATING ABM CRIME SIMULATION RESULTS

This section calculates several indicators for evaluating the simulated results by using the Kernel density estimation (KDE) based hot spot method. See Figure 13.4 for the workflow.

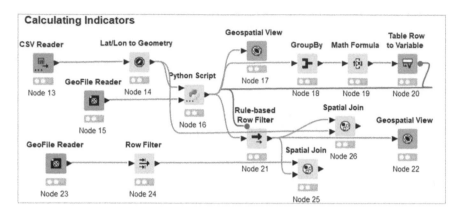

FIGURE 13.4 Workflow for Part 2 of Case Study 13

Step 5. Applying KDE analysis on simulated crimes: Add a CSV Reader node (Node 13) and input the file path for the simulation result table `sim_crime_inci-dents.csv` (e.g., `E:\CMS_GISV3\Chp13\ABMSIM\Data\sim_results\sim_crime_incidents.csv`). Remember to choose UTF-8 in the Encoding tab. The result table contains six columns, `OID_`, `Cycle`, `AgentID_Offender`, `AgentID_Target`, `X` and `Y`. The last two columns represent the location of simulated crime events.

Add a Lat/Lon to Geometry node (Node 14), and input `Y` for Latitude column, `X` for Longitude column, and `epsg:26915` for CRS. Click OK and execute it to transform the table as a spatial point data.

Add a GeoFile Reader node (Node 15) and input the file path for the boundary file `ebr_boundary.shp`, e.g., `E:\KNIME\CMGISV3\Chp13\data\ebr_bound-ary.shp`. Execute it to read the study area boundary.

Add a Python Script node (Node 16)[1] and configure it to have two input ports. In the dialog, copy the scripts from the `ABMKernel.py` to the coding box. Note that the values of `bandwidth` and `grid_size` are adjustable to control the smoothness of the density map or grid unit in meters. Click OK and execute it. The column density in the resulting table represents the KDE values, which can be visualized with a Geospatial View node (Node 17), as shown in Figure 13.5.

Step 6. Reclassifying the crime raster to generate hotspots: This step extracts hotspot areas based on the mean value and standard deviation of KDE values.

Add a GroupBy node (Node 18)[2] and link Python Script (Node 16) to it. In the dialog, (1) under the Groups tab, leave blank for Group column(s) and choose `Aggregation method (column name)` for Column naming, (2) under Manual Aggregation tab, choose `density` for Column, `Mean` and `Standard deviation` for Aggregation. Click OK and execute it. The resulting table has two columns: `Mean(density)` and `Standard deviation(density)` with their values 0.00072 (μ) and 0.00089 (σ), respectively.

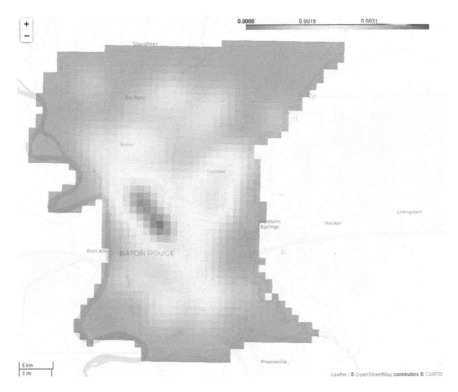

FIGURE 13.5   Kernel density estimation of simulated crime events

Here, we classify hotspots as any raster cell with two standard deviations above the average, i.e., $\mu + 2\sigma$. Add a Math Formula node (Node 19), and link GroupBy (Node 18) to it. Configure it to append a new column `Threshold` with the expression `$Mean(density)$+2*$Standard deviation(density)$`. The resulting `Threshold` column has a value of 0.0025.

Add a Table Row to Variable node (Node 20) and link Math Formula (Node 19) to it. Execute it directly to convert the three columns into flow variables. The Threshold variable will be used to filter hotspot units.

Add a Rule-based Row Filter node (Node 21), and link Python Script (Node 16) to its input data port and link Table Row to Variable (Node 20) to its input flow variable port. In the dialog, input `$density$ > $${DThreshold}$$=> TRUE` for the Expression and check Include TRUE matches. Click OK and execute it to extract the hotspot units. Use a Geospatial View node (Node 22) to visualize the extracted hotspot units, as shown in Figure 13.6.

Step 7. Calculating evaluation indices based on crime hotspots: In order to calibrate indices H, FPI, and FAI, we need to read the real crime data, i.e., 2011 reported robberies. Add a new GeoFile Reader node (node 23) and input the file path, e.g., `E:\KNIME\CMGISV3\Chp13\data\ebr_2011_robbery.shp`.

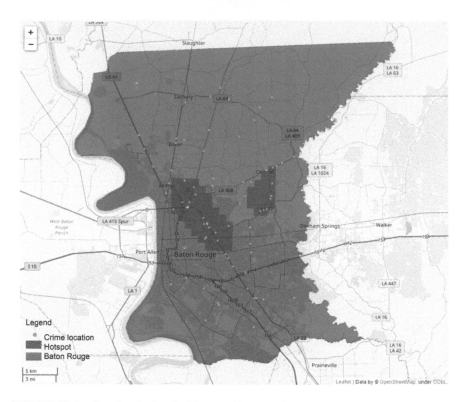

FIGURE 13.6 Simulated crime incidents and hotspots in the base scenario

Add a Row Filter node (Node 24) and link GeoFile Reader (Node 23) to it. In the dialog, (1) check Include rows by attribute value, choose OffenseMon for Column to test, (2) under Matching criteria, check use pattern matching, and choose 01. Click OK and execute it to extract robberies in January 2011. This identifies 76 robberies in that month ($N_r = 76$).

Add a Spatial Join node (Node 25) and link Rule-based Row Filter (Node 21) and Row Filter (Node 24) to it. Execute it directly with the default settings to get the reported crime inside the hotspot area extracted. 18 records fall within the predicted crime hotspots. Therefore, hit rate $H = \dfrac{18}{76}$ (= 23.68%).

To calculate *FPI* (Forecast Precision Index), we need the ratio of predicted hotspot size to that of study area (*a/A*). There are 1382 records (1KM × 1KM grids) in the result of Python Script (Node 16) as the study boundary, and 77 records (1KM × 1KM grids) in the result of Rule-based Row Filter (Node 21) as hotspots. Therefore, $a/A = 77/1382$, and $FPI = \dfrac{18/76}{77/1382} = 4.250$.

Add another Spatial Join node (Node 26), and link Row Filter (Node 25) and Lat/Lon to Geometry (Node 14) to it. Execute it directly with the default settings to get the reported crime inside the hotspot area extracted. There are 26 records falling

inside the predicted crime hotspots out of 77 simulated crime incidents (N_s = 29).
Therefore, $FAI = \dfrac{18/26}{76/77}$ (= 0.701).

This completes the simulation of crimes and evaluation of the model's goodness-of-fit in the base scenario. For each alternative scenario, repeat steps 3 to 7.

13.3 PART 3: TESTING VARIOUS HYPOTHESES BY ABM CRIME SIMULATION

First, we test the *journey-to-crime hypothesis*: with all other things being equal, a motivated offender is more likely to commit crime near its familiar locations because of better knowledge on target availability and avoiding apprehension. In other words, as the distance between an offender's home and current location increases, the probability for the offender to commit a robbery declines. A pre-calculated distance matrix between grid centers in a lookup table is utilized to avoid costly network-based distance calculation in real time. The result is compared to the base scenario with hotspot metrics.

Step 8. Simulating the scenario with offender journey-to-crime behavior: Double-click on the Component (Node 12) to open the dialog. In the dialog, select "True" for the "Apply journey-to-crime behavior" option (Figure 13.3). Click OK and re-execute the workflow.

After the simulation is completed, re-execute the workflow for calculating indicators (Nodes 13–26), which uses the updated sim_crime_incidents.csv from the previous workflow.

Another scenario is designed to test the effect of *neighborhood reputation on offender behavior*. The base scenario assumes that offender agents have no information on the reputation of neighborhoods and only use travel costs to select locations for flexible activities. Here, motivated offenders use both neighborhood reputation (e.g., more suitable targets) and travel cost to determine the probability of choosing a spatial grid for candidate locations. The total number of robberies inside each grid from the previous year (2010) is used as a proxy to measure the offender agent's knowledge of neighborhoods.

Step 9. Simulating the scenario with predefined crime opportunities: Similarly, double click the Component (Node 12) to bring out the dialog. Choose CRIME2010 for Crime opportunity field in spatial grid layer (Figure 13.3), then click OK and re-execute the workflow. The fields UTILITY and CRIME2010 comes from the data ebr_grid_500m_with_roads.shp. The field UTILITY has a constant value of 1 and assigns equal reputation to every grid in the base scenario. The field CRIME2010, on the other hand, records the total number of robberies in each grid in 2010.

After the simulation finishes, re-execute the workflow for calculating indictors (Nodes 13–26). The evaluation metrics indicate whether the inclusion of neighborhood reputation in offender decision-making process improves simulation performance in predicting crime hotspots.

The third scenario evaluates the effectiveness of police patrol strategies. In the base scenario, a police agent randomly travels to a patrol destination within its designated police district and stays there for a random period of up to 30 minutes. In the alternative scenario, a police agent uses the crime counts from the previous year to determine the probability of choosing a spatial grid for candidate patrol destinations. Grids with higher crime counts are favored by the *hotspot patrol strategy*.

Step 10. Exploring the scenario with hotspot police patrol: To configure this scenario, we need to update the value of the flow variable `police_hotspot_patrol` from `false` to `true`. This can be done either by using a Table Creator node, adding a new Single Selection Configuration node in the Component, or directly editing the value in the parameter CSV file using Excel or Notepad.

In addition, set the value of the variable `police_hotspot_field`[3] to `CRIME2010`. Once the updates are made, re-execute the entire workflow to generate the simulation scenario with the revised configuration.

This completes the simulations and evaluations for alternative scenarios for testing various hypotheses.

13.4   CONCLUDING REMARKS

The case study on the use of ABM crime simulation model examines how criminogenic features of local places and daily routines of vulnerable targets and motivated offenders affect spatiotemporal crime patterns. Various scenarios are simulated to test hypotheses suggested by crime theories and examine the impact of police patrol strategies.

In implementation, the case study describes the process of generating 59 model parameters in a predefined CSV file and a Table Column to Variable node. Local File Browser Configuration is used to generate full file paths for relative path parameters, while the nodes in Workflow Abstraction are utilized to create a component interface for adjusting parameters. The ABM simulation is executed using the Python Script node.

To calculate indicators, the workflow uses a Lat/Lon to Geometry node to convert simulated crime results into spatial points. A Python Script node is then used for Kernel density estimation to identify crime hotspots, while enlisting some help from common nodes of Spatial Join and GroupBy (mean and standard deviation).

NOTES

1   A similar node for Kernel Density estimation based on the Python script can be found in the latest version of Geospatial Analytics Extension for KNIME.
2   A Statistics node can also be used to calculate the mean and standard deviation.
3   Users can also assign police agents to different jurisdictions with separate configurations of patrol zone.

14 Spatiotemporal Big Data Analytics and Applications in Urban Studies

Where are the people in a city, at what time, and engaged in what activity? One major challenge in urban studies is to examine the spatiotemporal dynamics of human mobility in cities, especially in fine spatial and temporal resolutions. The availability of big data and rising computational power provide us an unprecedented opportunity to answer these questions in both spatial and temporal precisions quickly and effectively. The term "big data" is not necessarily defined by its size, rather emphasizes the sources other than surveys or mandatory reports or census. It is "found" or "observed" "passively as the digital exhausts of personal and commercial activities" (Census, 2022).

This chapter uses taxi trajectory data in Shanghai as an example to illustrate how to process and analyze spatiotemporal big data in KNIME.

The sample data set contains 900,862 records for taxis, i.e., about 0.8% of a larger taxi trajectory big data with more than 100 million locational records collected from more than 13,000 taxis running 24 hours in Shanghai. Each data entry contains 13 fields: Taxi ID, Alarm status, Vacant (Passenger Loading status), Roof light, Elevated Road, Brake, GPS receive time, GPS sent time, Longitude, Latitude, Speed, Heading, and Number of GPS Satellites.

The data folder `Shanghai` includes:

1. zipped Shapefile `Shanghai.zip` represents the administration boundary, and feature `SHroad.zip` is road network,
2. compressed CSV file `SHtaxi.csv.gz` includes sample taxi trajectory records, and
3. GeoPackage file `SHresult.gpkg` contains result files from Steps 3 and 4, layer `TOC` represents the taxi trips, and layers `TripOrigin` and `TripDestination` represent their origin and destination points, respectively.

14.1   CASE STUDY 14A: REBUILDING TAXI TRAJECTORY

This section constructs taxi trajectories through the road network from individual taxi trip records. Figure 14.1 shows the workflow of this case study.

DOI: 10.1201/9781003304357-14

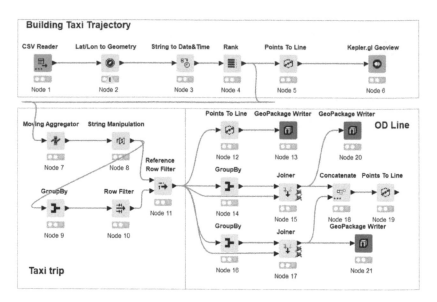

FIGURE 14.1 Workflow of Case Study 14A

Step 1. Visualizing taxi trajectories from spatiotemporal GPS-tracking points: This step uses the Points To Line tool in the Geospatial Analytics Extension to map taxi trajectories based on GPS tracking points, sorted by group and sort field.

In KNIME AP, create a new workflow Case14A. Add a CSV Reader node (Node 1) to the Editor and input the file path of sample taxi trajectory records in Node 1 (e.g., E:\KNIME\CMGISV3\Chp14\data\SHtaxi.csv), then execute it to read the data. Users can click Autodetect format in the dialog to adjust its reading settings.

Then, use a Lat/Lon to Geometry node (Node 2) to transform it as spatial point data. In the dialog, choose Latitude for Latitude column, Longitude for Longitude column and epsg:4326 for CRS. Click OK and execute it.

Add a String to Date&Time node (Node 3) and link Lat/Lon to Geometry (Node 2) to it. In the dialog shown in Figure 14.2, choose GPSreceive in the Include box, check Replace selected columns, keep Date&time for New type, yyyy-MM-dd HH:mm[:ss[.SSS]] for Date format, and keep other default settings. Click OK and execute it to convert the column GPSreceive into a time datatype column.

Add a Rank node (Node 4) and link String to Date&Time (Node 3) to it. In the dialogue shown in Figure 14.3, under Ranking Attributes tab, choosing TaxiID and GPSreceive for Column, and ascending for Order, and keep other default settings. Click OK and execute it to generate a column rank based on the order of TaxiID and GPSreceive.

Add a Points to Line node (Node 5) and link Rank (Node 4) to it. In the dialog shown in Figure 14.4, choose TaxiID for Group column, rank for Serial column, click OK and execute it. The resulting feature indicates the trajectories of individual taxis based on their IDs. With the help of Kepler.gl Geoview node (Node 6), the subset of taxi trajectory data reveals remarkable details of traffic intensity in Shanghai by visualization with transparency (Figure 14.5).

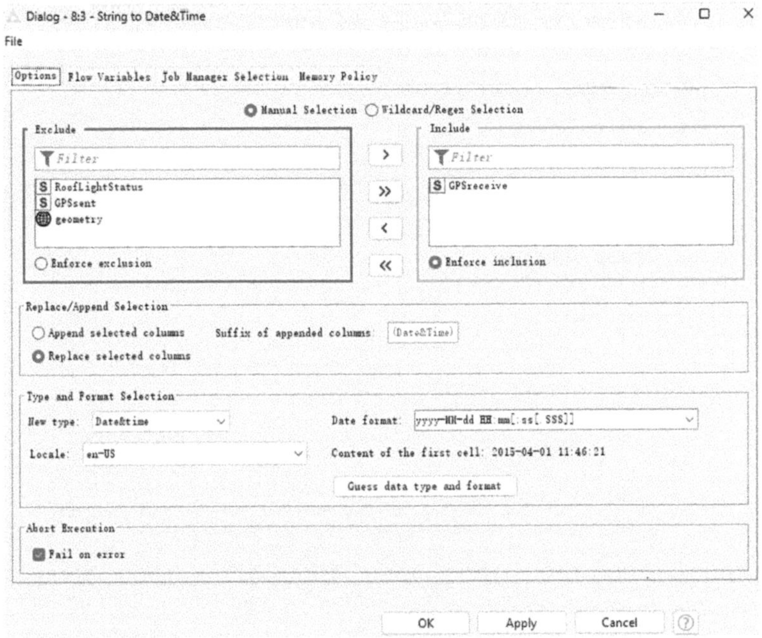

FIGURE 14.2   Interface of String to Date&Time

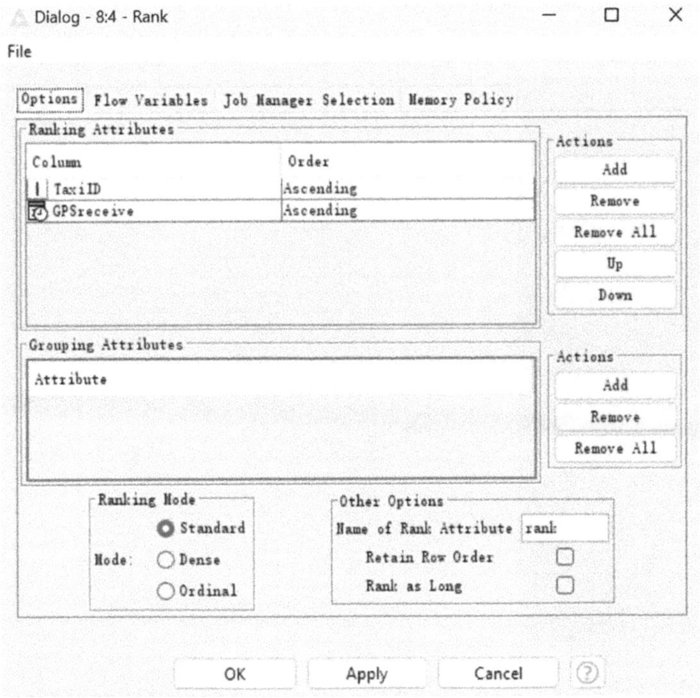

FIGURE 14.3   Interface of Rank

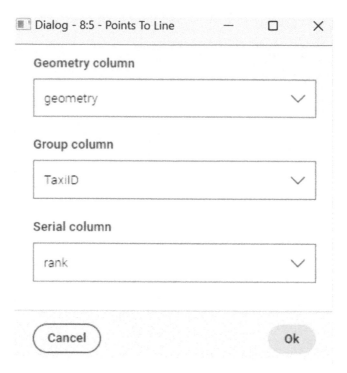

FIGURE 14.4 Interface of Points to Line

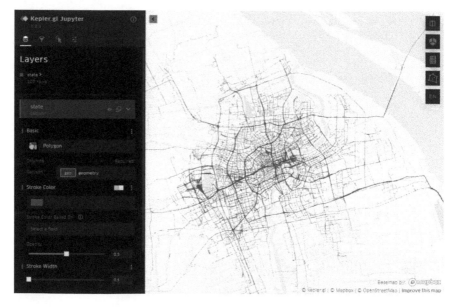

FIGURE 14.5 Taxi trajectories visualization

Step 2. Building tags TOC for trip vs. cruise: This step segments taxi trajectories into trips and cruises based on the values of column Vacant (0 for a trip, and 1 for a cruise).

Add a Moving Average node (Node 7) and link Rank (Node 4) to it. In the dialog, check Cumulative computation; under Aggregation Settings, choose Vacant for the column and Sum for the aggregation method. Execute the node to calculate the cumulative sum of Vacant, which identifies the start and end of a trip.

Add a String Manipulation node (Node 8) and link Moving Average (Node 7) to it. Configure it to append a new column TOC with the expression join-Sep("_",string($TaxiID$),string($Sum(Vacant)$)). This expression creates a new column that concatenates the TaxiID and the cumulative sum of the column Vacant, separated by an underscore.

The records sharing the same value in the column TOC are the component points of an entire taxi trip, such as all the records with a TOC label of 10461_0. The records with ascending TOC labels (e.g., 10490_5113, 10490_5114, etc.) are the points of an entire taxi cruise.

By segmenting the trajectories into trips and cruises, we can better understand the spatial connections between places and analyze the patterns of taxi movements.

Step 3. Constructing trips from points by the TOC label: Add a GroupBy node (Node 9) to summarize the trip counts marked by the column TOC, and link String Manipulation node (Node 8) to it. In the dialog, (1) under the Groups tab, choose TOC for Group column(s), Aggregation method(column name) for Column naming, (2) under Manual Aggregation tab, choose TaxiID for Column and Count for Aggregation. Click OK and execute it. The column Count *(TaxiID) represents the total number of trips for each unique trip ID in the column TOC.

Add a Row Filter node (Node 10) and link GroupBy (Node 9) to it. Configure it to filter out the records with Count *(TaxiID) < 18. This removes short trips and cruises with only one record. Recall that the GPS receiver interval is 10 seconds, which means that filtering out records not only eliminates cruising records (Count*(TaxiID) = 1), but also short trips lasting only 3 minutes. An alternative option is to set the distance threshold to 200 after using Points To Line node to convert it into lines and calculating length by the Length node, which would identify short paths resulting from erroneous data and filter out very short trips (e.g., < 0.2 km).

Add a Reference Row Filter node (Node 11) and link Row Filter (Node 10) to it. In the dialog, choose TOC for both Data column and Reference column, check Include, click OK and execute it. The output table contains all the points of trips with their unique trip IDs in the column TOC.

Add another Points To Line node (Node 12) and link Reference Row Filter (Node 11) to it. In the dialog, choose TOC for Group column, rank for Serial column, click OK and execute it. The resulting table shows the trips with their origins and destinations, segmented based on the Vacant column (trip vs. cruise) and removing short trips and cruises with only one record. There are a total of n = 3,842 trips.

Add a GeoPackage Writer Node (Node 13) and link Points To Line (Node 12) to it. Configure the node to save the data as a layer TOC in a GeoPackage file, e.g., E:\KNIME\CMGISV3\Chp14\data\SHresult.gpkg.

Step 4. Aggregating numbers of trips by origins and destinations: In order to create an OD (Origin-Destination) list, we need to identify the start and end points of each trip. This is accomplished by summarizing the minimum and maximum values of column `rank`.

First, add a GroupBy node (Node 14) and link Reference Row Filter (Node 11) to it. Configure it to summarize the `Minimum` value of rank based on the Group column `TOC`, be sure to choose `Keep original name(s)` for Column naming in the dialog. Next, add a Joiner node (Node 15), and link GroupBy (Node 14) and Reference Row Filter (Node 11) to it. In the dialog, (1) under Joiner Settings tab, choose `TOC` and `rank` for both Top Input ('left' table) and Bottom Input ('right' table); (2) under Column Selection, exclude `TOC` and `rank` from any of the Include box to avoid duplicated column names, keep other default settings and execute it. This step extracts the start points for all trips.

Copy GroupBy (Node 14) and Joiner (Node 15) as two new nodes (Nodes 16 and 17). Repeat the same linkage (Node 11→ Node 16, Node 16 and Node 11→ Node 17) and use the same parameter settings except change `Minimum` to `Maximum` in the dialog of GroupBy (Node 16) to extract the end points.

Add a Concatenate node (Node 18) and link the two Joiner nodes (Nodes 15 and 17) to it. Execute it to combine these two data vertically. Finally, copy Points To Line node (Node 12) as a new node (Node 19) and link Concatenate (Node 18) to it. Execute it with the predefined parameters in the node to generate the OD straight lines derived from the taxi trip trajectories.

Add two GeoPackage Writer Nodes (Nodes 20 and 21), link Joiner (Node 15) to Node 20, and link Joiner (Node 17) to Node 21. In the dialog shown in Figure 14.6, configure

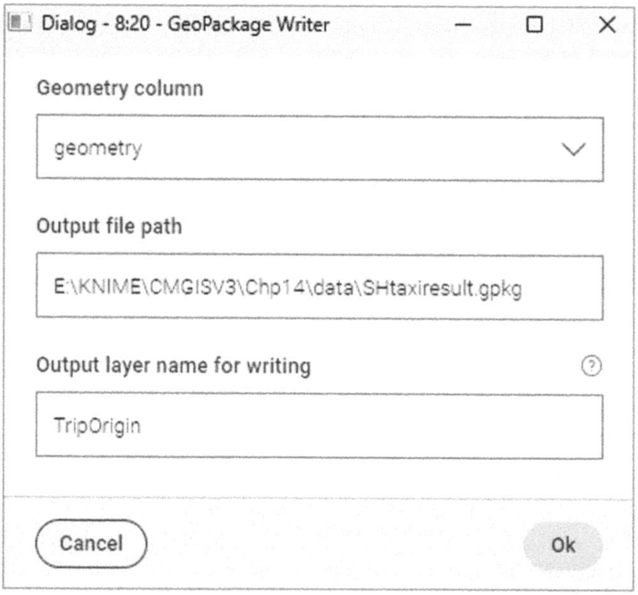

FIGURE 14.6   Interface of GeoPackage Writer

the two nodes to save the data as two layers `TripOrigin` and `TripDestination` in the previous file `SHresult.gpkg`, respectively.

14.2   CASE STUDY 14B: AGGREGATING TAXI TRAJECTORY BETWEEN GRIDS

This section aggregates taxi trajectories to O-D patterns between uniform grids. Figure 14.7 shows the workflow of this case study.

Step 1. Aggregating interaction volumes between grids: In KNIME AP, create a new workflow `Case14B`. Add a GeoPackage Reader node (Node 1) to read the layer `TOC` in the file `SHresult.gpkg`, generated in Node 12 in `Case14A`. Add a Projection node (Node 2) and link GeoPackage Reader (Node 1) to it. In the dialog, input `epsg:3857` for New CRS, check `Append` for Output column. Execute it to convert it to a projected coordinates system. Add a Create Grid node (Node 3) and link Projection (Node 2) to it. In the dialog, input `1000` for Grid length. Execute it to create a 1km×1km grid that covers the extent of all taxi trajectories.

Add two GeoPackage Reader nodes (Nodes 4 and 5) to read the layers `TripOrigin` and `TripDestination` in the file `SHresult.gpkg`, generated in Nodes 15 and 17 in `Case14A`, respectively. They are the origin and destination points of all taxi trips.

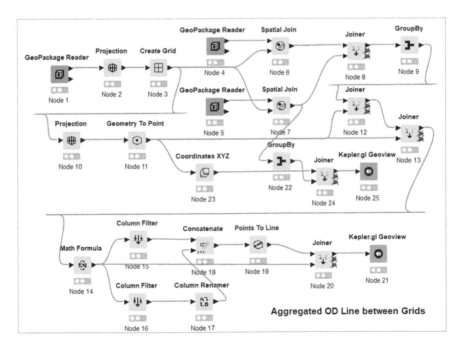

FIGURE 14.7   Workflow of Case Study 14B

Add two Spatial Join nodes (Nodes 6 and 7), link GeoPackage Reader (Node 4) and Create Grid (Node 3) to Spatial Join (Node 6), and connect GeoPackage Reader (Node 5) and Create Grid (Node 3) to Spatial Join (Nodes 7). Execute the two nodes with the default settings to join `Grid ID` to the start points or end points that fall into the corresponding grid.

Add a Joiner node (Node 8), and link the two Spatial Join nodes (Nodes 6 and 7) to it. In the dialog, (1) under Joiner Settings tab, choose `TOC` for both Top Input ('left' table) and Bottom Input ('right' table); (2) under Column Selection, keep `TOC` and `Grid ID` for the Include box of Top Input (left table), `Grid ID` for the Bottom Input (right table), and (3) keep other default settings. Execute it to join the columns, `Grid ID` for origin and `Grid ID (right)` for destination.

Add a GroupBy node (Node 9) and link Joiner node (Node 8) to it. In the dialog, (1) under the Groups tab, choose `Grid ID` and `Grid ID (right)` for Group column(s), `Aggregation method(column name)` for Column naming, (2) under Manual Aggregation tab, choose `TOC` for Column and `Count` for Aggregation. Click OK and execute it. In the result table, the field `Count*(TOC)` represents the number of trips between grids.

Step 2. Building OD trips between grids: To rebuild the trips among the grids, we need to attach the grid centroids to the OD list. Add a Projection node (Node 10) and link Create Grid (Node 3) to it. Configure it to convert the data into a geographic coordinates system (`epsg: 4326`). Then, use a Geometry To Point node (Node 11) to convert the grids into their centroids.

Next, add two Joiner nodes (Nodes 12 and 13), link GroupBy (Node 9) and Geometry To Point (Node 11) to Joiner (Node 12), and link Joiner (Node 12) and Geometry To Point (Node 11) to Joiner (Node 13). In the dialog of Joiner (Node 12), (1) under Joiner Settings tab, choose `Grid ID` for both Top Input ('left' table) and Bottom Input ('right' table); (2) under Column Selection, keep all columns in the Include box of Top Input (left table), keep `geometry` for the Bottom Input (right table), keep other default settings and execute it. This step joins the grid centroids as the origin points.

Joiner (Node 13) uses the same setting as Node 12, except choosing `Grid ID (right)` for Top Input ('left' table) and `Grid ID` for Bottom Input ('right' table) under Joiner Settings tab. This step joins the grid centroids as the destination points, which is named as `geometry(#1)` to avoid duplicated names.

Add a Math Formula node (Node 14) and link Joiner (Node 13) to it. Configure it to append a new integer column `ID` with the expression `$$ROWINDEX$$+1`.

Add two Column Filter nodes (Nodes 15 and 16), and link Math Formula (Node 14) to them. Use Node 15 to keep the two columns `ID` and `geometry,` and use Node 16 to keep the two columns `ID` and `geometry(#1)`. Before combining the two tables, we can use a Column Renamer node (Node 17) and link Column Filter (Node 16) to it. Configure it to rename `geometry(#1)` as `geometry`. Add a Concatenate node (Node 18), link Column Filter (Node 15) and Column Renamer (Node 17) to it. Execute it directly to combine the two tables vertically.

Add a Points To Line node (Node 19), and link Concatenate node (Node 18) to it. In the dialog of Node 19, choose `ID` for both Group column and Serial column.

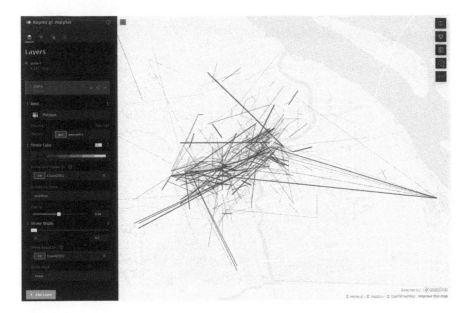

FIGURE 14.8   Aggregated trips between 1km × 1km grids

Execute it to generate the trips between grids. Add a Joiner node (Node 20), and link Points To Line (Node 19) and Math Formula (Node 14) to it. In the dialog, (1) under Joiner Settings tab, choose ID for both Top Input ('left' table) and Bottom Input ('right' table); (2) under Column Selection, keep all columns in the Include box of Top Input (left table), and keep Count*(TOC) for the Bottom Input (right table); (3) keep other default settings, click OK and execute it.

This step links the trip volumes to the line geometry data. The result can be visualized by a Kepler.gl Geoview node (Node 21), as shown in Figure 14.8.

Step 3. Visualizing destination heatmap: This step leverages additional options available in the Kepler.gl Geoview node to generate the heatmap. The task begins with attaching coordinates to the OD trips.

Add a GroupBy node (Node 22) and link Spatial Join (Node 7) to it. In the dialog, (1) under the Groups tab, choose Grid ID for Group column (s) and Aggregation method(column name) for Column naming, (2) under Manual Aggregation tab, choose TOC for Column and Count for Aggregation. Click OK and execute it. The column Count*(TOC) represents the total number of times a trip's destination falls in a grid.

Add a Coordinates XYZ node (Node 23) and link Geometry To Point (Node 11) to it. Execute it to append two columns for coordinates, x and y, representing longitude and latitude, respectively.

Add a Joiner node (Node 24), and link Coordinates XYZ (Node 23) and GroupBy (Node 22) to it. In the dialog, (1) under Joiner Settings tab, choose Grid ID for both Top Input ('left' table) and Bottom Input ('right' table); (2) under Column Selection,

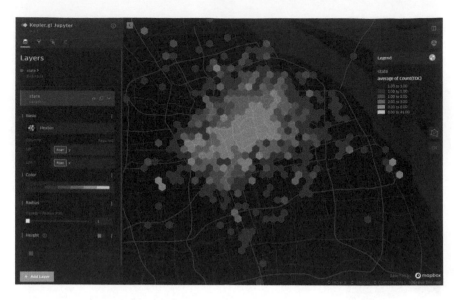

FIGURE 14.9 Destination heatmap

keep all columns in the Include box of Top Input (left table), keep `Count*(TOC)` for the Bottom Input (right table); (3) keep other default settings and execute it.

Use a Kepler.gl Geoview node (Node 25) to visualize the output setting `Count*(TOC)` as Stroke Width, as shown in Figure 14.9. The map captures the variability of intensity for places being visited.

14.3 CONCLUDING REMARKS

This chapter discusses the challenge of analyzing spatiotemporal dynamics of human mobility in urban areas and how geospatial big data can be used to address this challenge. The case study uses a scaled-down taxi trajectory dataset in Shanghai to illustrate the process of big data analysis in urban studies. The dataset contains over 900,000 records and includes information on taxi ID, location, and other relevant details.

The Points To Line node in the Geospatial Analytics Extension is used to reconstruct trajectories based on time and ID tags in the workflow. The Moving Average node is useful in distinguishing between trips and cruising in taxi trajectory data. The GroupBy node exports the origin or destination points by aggregating the minimum or maximum time tag for a taxi, with the support of the Rank node.

Given the nature of taxi data, or big data in general, at the finest spatial resolution (i.e., individual points), we use a technique similar to Monte Carlo simulation in Chapter 12 to aggregate the data of points (e.g., trip ends) and links (e.g., OD trips) in an area unit (here grids of uniform size) to examine spatial patterns. The difference is: Chapter 12 used simulated individuals from observed aggregated patterns to gain spatial accuracy and this chapter is based on individuals in the real world.

To work with the complete and much larger dataset, users only need to replace the data in the CSV Reader node, and integrate the two workflows by copying all nodes from Case14B to Case14A and use Point To Line (Node 12), Joiner (Node 15), and Joiner (Node 17) in Case14A to replace the links of the three GeoPackage Reader nodes in Case14B, respectively.

References

Census, U.S. Bureau. 2022. Big Data. https://www.census.gov/topics/research/big-data. html#:~:text=Big%20data%20is%20a%20term,%2C%20and%20%20%22smart%22%20 devices.

Gu, C., F. Wang, and G. Liu. 2005. The structure of social space in Beijing in 1998: A socialist city in transition. *Urban Geography*, 26(2), 167–192.

Hu, Y., F. Wang, C. Guin, and H. Zhu. 2018. A spatio-temporal kernel density estimation framework for predictive crime hotspot mapping and evaluation. *Applied geography*, 99, 89–97.

Louisiana Hospital Association. 2021. Louisiana Hospital Association Membership Directory. Available https://cdn.ymaws.com/www.lhaonline.org/resource/resmgr/about/2020_ 21LHADirectory.pdf.

Luo, J., L. Tian, L. Luo, H. Yi, and F. Wang. 2017. Two-step optimization for spatial accessibility improvement: A case study of health care planning in rural China. *BioMed Research International* 2017, 2094654

Mu, L., F. Wang, V. W. Chen, and X. C. Wu. 2015. A place-oriented, mixed-level regionalization method for constructing geographic areas in health data dissemination and analysis. *Annals of the Association of American Geographers*, 105(1), 48–66.

Schroeder, J. P. 2007. Target-density weighting interpolation and uncertainty evaluation for temporal analysis of census data. *Geographical Analysis* 39, 311–335.

Wang, C., and F. Wang. 2022. GIS-automated delineation of hospital service areas in Florida: From Dartmouth method to network community detection methods. *Annals of GIS*, 28(2), 93–109.

Wang, F. 1998. Urban population distribution with various road networks: a simulation approach. *Environment and Planning B-Planning & Design* 25: 265–278.

Wang, F. 2000. Modeling commuting patterns in Chicago in a GIS environment: A job accessibility perspective. *The Professional Geographer*, 52(1), 120–133.

Wang, F. 2001. Explaining intraurban variations of commuting by job accessibility and workers' characteristics. *Environment and Planning B: Planning and Design* 28, 169–182.

Wang, F. 2003. Job proximity and accessibility for workers of various wage groups. *Urban Geography* 24: 253–271.

Wang, F., D. Guo and S. McLafferty. 2012. Constructing geographic areas for cancer data analysis: a case study on late-stage breast cancer risk in Illinois. *Applied Geography* 35: 1–11.

Wang, F., Y. Hu, S. Wang, and X. Li. 2017. Local indicator of colocation quotient with a statistical significance test: examining spatial association of crime and facilities. *The Professional Geographer*, 69(1), 22–31.

Wang, F., C. Liu, and Y. Xu. 2019. Mitigating the zonal effect in modeling urban population density functions by Monte Carlo simulation. *Environment and Planning B: Urban Analytics and City Science*, 46(6), 1061–1078.

Wang, F., M. Vingiello, and I.M. Xierali. 2020. Serving a Segregated Metropolitan Area: Disparities in Spatial Access to Primary Care Physicians in Baton Rouge, Louisiana. In: Lu, Y., Delmelle, E. (eds) Geospatial Technologies for Urban Health. Global Perspectives on Health Geography. Springer, Cham. https://doi.org/10.1007/978-3-030-19573-1_5

Wang, F., G. Wang, J.F. Hartmann and W. Luo. 2012. Sinification of Zhuang place names in Guangxi, China: A GIS-based spatial analysis approach. *Transactions of the Institute of British Geographers* 37: 317–333.

Wang, F., and Y. Xu. 2011. Estimating O-D matrix of travel time by google maps API: Implementation, advantages and implications. *Annals of GIS 17*, 199–209.

Wang, F., L. Zhang, G. Zhang, and H. Zhang. 2014. Mapping and spatial analysis of multiethnic toponyms in Yunnan, China. *Cartography and Geographic Information Science*, *41*(1), 86–99.

Index

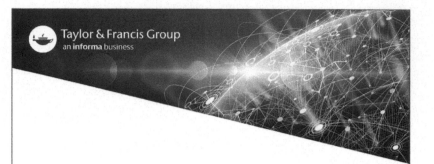